T0348549

Guide to
Wear Problems and Testing for Industry

Guide to

Wear Problems and Testing for Industry

M J Neale and M Gee

William Andrew Incorporated
Norwich, New York, USA

First published in 2000 by Professional Engineering Publishing Limited, UK

ISBN 0-8155-1471-9
Library of Congress Card Number 00-108631

This edition distributed exclusively by:
William Andrew Publishing, LLC
13 Eaton Avenue
Norwich, NY 13815, USA
Phone: (800) 932 7045 or (607) 337 5080
Fax: (607) 337 5090
E-mail: sales@williamandrew.com
http://www.williamandrew.com

Printed and bound in the United Kingdom

Transferred to Digital Printing, 2011

Contents

Chapter 1

INTRODUCTION

Manufacturing industry and public utilities own and operate a large amount of plant and machinery. The machinery may be large in size as in power stations and steel works, or it may be relatively small as in electronic manufacture. All these machines are prone to wear, either of their own relatively moving parts, or where they come into moving contact with various materials. A national survey in 1997 has indicated that the cost of this wear to UK industry was of the order of £650 million per year. Also, for companies who have these wear problems, the costs were typically about 0.25 percent of their turnover. In many cases, these costs can be at least halved by making appropriate design and/or material changes, to reduce or eliminate wear.

Changes to the design of plant and machinery are usually seen as rather major alterations by the owners and users, with an associated risk. In contrast it is relatively easy to use improved materials when fitting replacement parts, but this leaves the problem of identifying what is an improved material. There is, therefore a need to have information on the relative wear performance of various materials in typical industrial applications.

This guide reviews the wear mechanisms that occur in the various types of industrial plant and machinery, and gives general guidance on the relative wear performance of the different materials which can be used for their components. It also reviews the laboratory tests that are available for simulating the practical wear conditions, to give a comparison of the expected wear performance of various candidate replacement materials. This testing provides a relatively cheap, safe, and quick procedure compared with experimenting with components made from various materials in operating service.

Chapter 2

INDUSTRIAL WEAR PROBLEMS

Wear is a process in which material is removed from the surfaces of components, or by which these surfaces are seriously disturbed. In order to reduce wear, it is important to understand the mechanism by which it occurs in each case. There are a number of different types of wear and each one requires a different practical approach to wear reduction.

2.1 Types of wear

The seven main types of wear are shown diagramatically in Table 2.1 (on pages 6 and 7), and are described in more detail in the following paragraphs.

Abrasive wear occurs when material is removed from the surface of a component by a cutting action. This may be an intended and controlled process in component manufacture, such as filing or grinding, or it may occur randomly in machine operation, such as the wear of digger teeth when working in gravel. It can also occur with two smooth surfaces rubbing together but with small hard contaminant particles trapped in between them. Abraded surfaces show damage which can range from fine scratching to deep gouges. If the component is made from a ductile material, such as steel, the wear debris can be spiral in shape, similar to machining swarf. For very hard materials the wear debris tends to be in the form of chips, generated by local brittle fracture of the material.

Adhesive wear is the surface damage and material removal which can occur when two smooth surfaces rub against each other. Such surfaces are never perfectly smooth and have high spots where the rubbing occurs. These local areas experience concentrated contact loads and interactions, and tend to adhere to each other and drag material away along the surface. This type of wear can occur in plain bearings and other interacting machine components, particularly if they are inadequately lubricated. The wear of brakes and clutches occurs by the same mechanism, and is kept under control by the use of dissimilar materials. Surfaces subject to adhesive wear can end up polished, with the generation of fine flakes of wear debris, or can show severe surface damage associated with surface dragging or even seizure.

Fretting is a particular form of adhesive wear which occurs when there are small oscillatory movements between two surfaces. These movements can arise from the deflection of machine components with clamped joints or press fits, or they may be intended small movements as in gear couplings. Fretting often produces fine powdered and oxidized wear debris around the components and there is usually surface damage and roughening of the joint surfaces. Fretting is also a limiting factor in the life of wire ropes. The individual steel wires rub against each other due to relative deflections when the load on the rope changes or when it is bent around a pulley. The oxidized wear debris often produces a red staining on the surface of the rope.

Erosion involves the removal of material from the surface of a component by the high-speed impact of a liquid or of a stream of hard particles carried in a fluid flow. The two common types of erosion are cavitation erosion and particle erosion. Cavitation erosion occurs on components subject to low transient fluid pressures, such as ships' propellers, and arises from the intense local impact of the collapse of low-pressure vapour bubbles on to the component surface. Particle erosion occurs when a stream of hard particles is directed at a surface. This may be intended as in shot blasting processes, or it may arise incidentally, such as in pipelines and associated components carrying slurries, or crude oil containing sand.

Fatigue of surfaces can also lead to the loss of material when fatigue cracks in the surface join together to create loose particles. This surface fatigue can arise from a contact stress fatigue mechanism or from a thermal stress fatigue mechanism. Contact fatigue can arise in rolling contact where the passage of a ball or roller over a surface causes alternating tensile and compressive stresses, which can create fatigue cracks. Thermal fatigue cracking can arise from transient heating and cooling of a surface, particularly when combined with surface frictional forces as in clutch plates and heavily loaded plain bearings. It can be particularly severe if a surface is in intermittent contact with a very hot material, such as molten metal.

Some of these types of wear cause more problems than others and give rise to higher costs in the operation of machinery and plant. Table 2.2 shows the experience of industry in the UK in terms of the relative costs arising from the different kinds of wear.

Table 2.1
The types of wear that occur in industrial machines

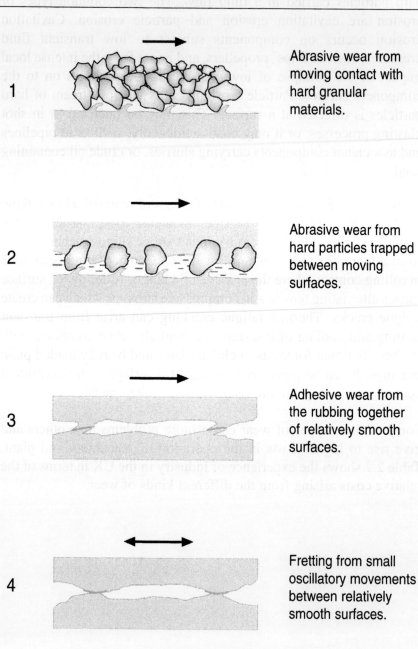

1. Abrasive wear from moving contact with hard granular materials.

2. Abrasive wear from hard particles trapped between moving surfaces.

3. Adhesive wear from the rubbing together of relatively smooth surfaces.

4. Fretting from small oscillatory movements between relatively smooth surfaces.

Table 2.1 (continued)
The types of wear that occur in industrial machines

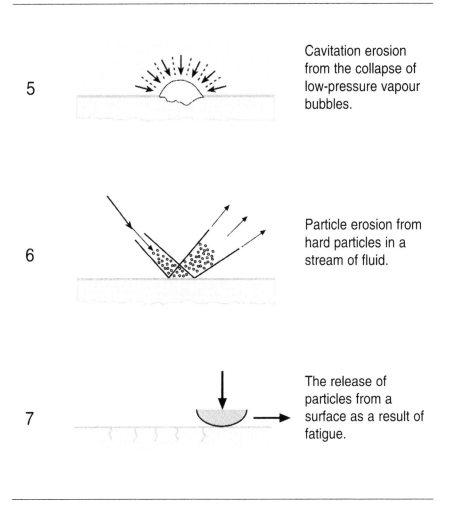

5	Cavitation erosion from the collapse of low-pressure vapour bubbles.
6	Particle erosion from hard particles in a stream of fluid.
7	The release of particles from a surface as a result of fatigue.

Table 2.2
Relative importance of the various types of wear

Type of wear	Approximate percentage contribution to the cost of wear (%)
Abrasive wear	63
Adhesive wear and fretting	26
Erosion and surface fatigue	11

2.2   Identification of the type of wear

The seven most significant types of wear have been listed in Table 2.1 with the mechanisms that give rise to them. To solve an industrial wear problem, it is useful to know which type (or types) of wear is occurring. This can generally be identified from:

1. *The appearance of the worn surfaces*   Typical examples of surfaces which have experienced the various types of wear are shown in Figs 2.1 to 2.7. [The authors acknowledge the provision of illustrations given in Figs 2.1 to 2.7 from Dr Terry Eyre (Eyre Consultants) and Dr Robert Wood (University of Southampton).] Guidance on methods of examining worn surfaces is given in Chapter 5.

2. *The nature of the application*   For example, significant abrasive wear will only occur in the presence of hard granular materials, and erosion will only occur in the presence of fluids with a high velocity relative to the surface. Table 2.3 (on page 16) gives some examples of industrial applications in which the various types of wear commonly occur. It is desirable to analyse the practical operating conditions in considerable detail to try to identify the combination of circumstances that are causing the wear. These circumstances may be transient or cyclic and a detailed analysis

of the loading and movement patterns, contact conditions and operating environment can provide useful guidance both for identifying the wear mechanism and for planning tests to simulate it.

Wear, however, is not a simple process, and while the individual mechanisms can be described, it needs to be remembered that there can be transitions between one type and another. For example, in an adhesive wear situation, surfaces can harden due to working or thermal effects, and these harder materials can then cause some abrasion.

It is also very common for more than one mechanism of wear or material removal to occur simultaneously. For example corrosion often interacts with abrasive wear to give increased wear rates.

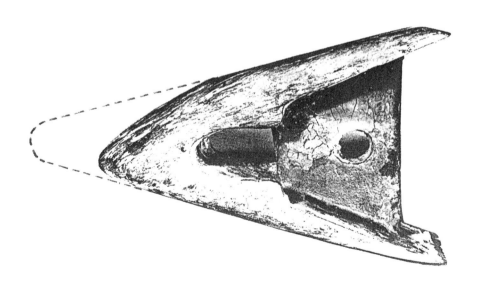

**Fig. 2.1 Abrasive wear from moving contact
with hard granular materials**

*Fig. 2.2 Abrasive wear from hard particles trapped
between moving surfaces*

*Fig. 2.3 Adhesive wear from the rubbing together of
relatively smooth surfaces*

***Fig. 2.4 Fretting damage on a surface from small oscillating
movements against a mating surface***

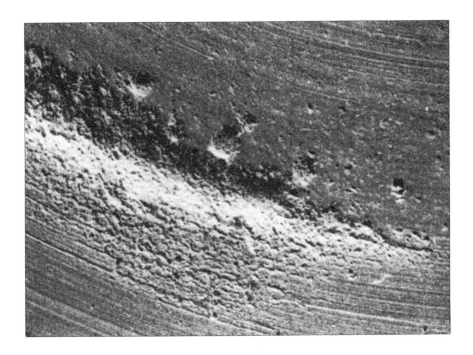

***Fig. 2.5 Cavitation erosion from the collapse
of low-pressure vapour bubbles***

Fig. 2.6 Erosion by hard particles in a high-velocity fluid stream

Contact stress fatigue

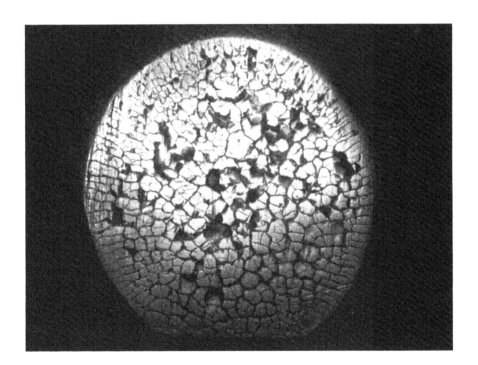

Thermal fatigue

Fig. 2.7 Surface fatigue cracking which allows the release of large particles

Table 2.3
Practical examples of the occurrence of the various types of wear

Type of wear	Examples of industrial applications where this wear is likely to occur	
1. Abrasive wear from contact with hard granular material	Earth-moving machines Material-handling chutes Rock crushers Mining conveyors	Rotors of powder mixers Extrusion dies for bricks and tiles
2. Abrasive wear from hard particles trapped between moving surfaces	Pivot pins in construction machinery Scraper blades in plaster-mixing machines	
3. Adhesive wear from the rubbing together of smooth surfaces	Rubbing bearings Clutches Press tools	Brakes Piston rings and cylinder liners
4. Fretting	Connecting rod joints in internal combustion engines Bearing-to-housing contact, with dynamic loads Spline and gear couplings with misalignment Wire ropes and overhead electrical conductor cables	
5. Cavitation erosion	Marine propellers Pump rotors Hydraulic control valves	
6. Particle erosion	Valves controlling the flow of sandy crude oil Pipelines carrying abrasive materials Helicopter rotors in desert operation	
7. Surface fatigue	Rolling bearing races and elements Heavily loaded high-speed plain bearings Clutches subject to excessive slip Surfaces in intermittent contact with molten metals	

2.3 Assessing the problem

Wear is a common occurrence on most plant and machinery, and is often a slow and progressive process, which may be accepted as normal. However, if the rate of wear of a particular machine component is high, so that it requires frequent repair and replacement, then it may constitute a wear problem. Therefore, deciding whether a wear problem exists and requires attention calls for a degree of judgement of the circumstances. For example, if there is any risk of an accident arising from the wear, or any major customer dissatisfaction, this would require immediate action.

Guidance on the existence of a wear problem can also be obtained from an analysis of the contribution that it makes to the operating cost of a piece of plant or machinery. A survey of wear problems and costs in UK industry in 1997 indicated that a typical average cost of wear was about 0.25 percent of company turnover. If, therefore, in a particular plant, its cost is appreciably above this percentage of the output value, this suggests that wear may be a problem justifying further attention.

An alternative method of assessment, is to compare the rate of wear that is occurring on a particular machine with typical values on similar equipment. This is a very relevant method if the machine is being sold as a product, since its performance can then be compared in the market place with that of competitive products. Also if wear is the major factor limiting the life of a particular type of product, there will be a major market advantage available to a manufacturer who can produce a product with an improved wear performance. It may be possible to construct or obtain diagrams of the wear performance of various types of components or machines, such as the data in Fig. 2.8 for the wear performance of the cylinders of internal combustion engines. Such diagrams are likely to change with time as a result of technical development. Figure 2.8 indicates the wear rates being obtained towards the end of the twentieth century. A hundred years earlier, when internal combustion engines were first being developed, cylinder liner wear rates were much higher. They have since been reduced by the use of better materials and designs, and particularly by changes in the environment towards higher temperatures, to eliminate condensation and cylinder corrosion.

It also needs to be remembered that problems due to wear may show themselves in ways other than changes in component dimensions. For example, problems may arise from the generation of wear particles, which, if they cannot escape from the operating area, may cause increased friction or seizure. They may also cause problems with the blocking of oil filters in lubricated applications.

On most machines which have a wear problem, a common first approach is to try a change in the material or surface condition of the component involved. If the components are lubricated, a change in the lubricant may also be tried. This is a simple first approach to the problem compared with trying a more major design change. It does, however, require guidance on the selection of appropriate alternative materials, which in turn can be given further support by appropriate wear testing.

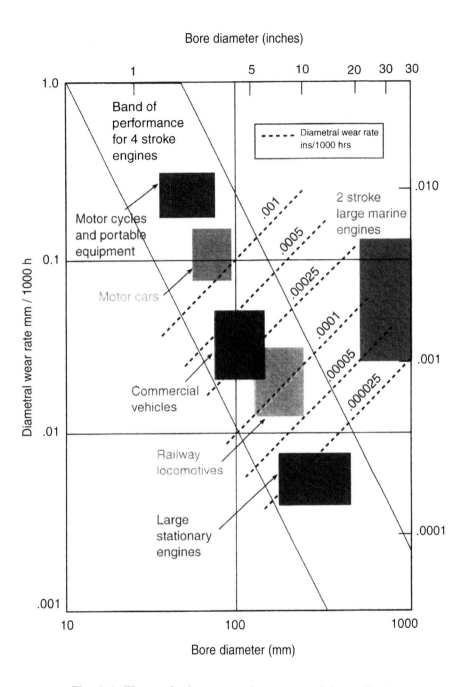

Fig. 2.8 The typical wear performance of the cylinders of internal combustion engines (to see this figure in colour, please refer to colour plate section)

2.4   Improvement by the use of alternative materials

The seven main types of wear which occur in plant and machinery have been listed in Tables 2.1 and 2.3 and the appearance of components experiencing these types of wear has been illustrated in Figs 2.1 to 2.7, to assist in their identification.

The basic wear mechanisms underlying the different types of wear are generally understood and can be used as guidance to the choice of appropriate wear resisting materials. This approach is used in the following sections of this chapter to look in turn at the seven types of wear, and the applications where they occur, to provide initial guidance on the selection of improved materials. This can help to provide a short-list of materials for appropriate laboratory wear testing, or with an appropriate balance of confidence, need, and risk, can be used to determine materials to be tried directly in the practical application.

2.4.1 Abrasive wear by hard, granular materials

This type of wear occurs on the blades and buckets of earth-moving machinery and on production plant handling abrasive materials, such as chutes and conveyors, extrusion dies for bricks and tiles, and the rotors of powder mixers.

In this type of wear, the sharp edges of hard particulate material can cut into the component surface, which then shows damage ranging from fine grooving to gouging. If the hard particles are more rounded they can slide or roll along the surface and create grooves caused by material displacement rather than cutting. This plastic deformation can create lips at the sides of the grooves, which can then fracture away and create wear debris. If the wear is by a bulk mass of abrasive material, the amount of wear and the wear patterns can also be affected by the degree of compliance of the material behind the surface layers of abrasive particles. If the support is very compliant, as in sand/clay mixtures, the contact loads at the abrasive particles become more equally distributed, which can make the wear less severe.

The wear rate of the components can be reduced by making them harder. However, very hard materials tend to be brittle, and components made from them tend not to be robust. In many cases, there is little improvement in wear resistance if the hardness of the component exceeds 1.3 times the hardness of the abrasive, and a compromise between hardness and toughness is generally chosen. The abrasive wear resistance of ferrous materials increases with carbon contents up to 3 percent, with the carbon in the form of cementite. White cast irons with carbon contents of this order are widely used for abrasive wear resistance.

Austenitic manganese steels with manganese content up to 14 percent are also very effective in severe abrasive conditions, because they work harden with heavy abrasive impacts. For this reason they are, however, difficult to machine and are therefore generally used as castings.

Very good wear resistance with robustness can also be achieved by applying a hard surface coating to a robust core material. Tungsten carbide cobalt composites can be brazed to the surface of steel components. Other hard materials such as chromium can be applied by plating, carbides by powder spraying, and martensitic irons by welding. It is important in this situation to use a hard coating of adequate thickness. If the coating is too thin, the underlying core material can plastically deform, under the contact loads, and allow the hard coating to crack off, like an eggshell.

In wet environments, corrosion can increase the rate of wear because abrasive action can remove passivated films created on the surface by corrosion and expose fresh metal, which is more prone to further corrosion than the passivated surface. If a corrosive fluid is present this effect is increased and the surface corrosion can also stimulate the formation of fatigue cracks in the surface. In all these circumstances, the corrosion resistance of the material then becomes a significant factor in its wear performance.

2.4.2 *Abrasive wear from hard particles trapped between moving surfaces*

This type of wear occurs on exposed pivot pins and bushes operating in an abrasive environment, such as linkage pins on earth-moving machines. Similar conditions also exist under the scraper blades of plaster-mixing machines, and in other machines handling abrasive materials.

In principle, the wear could be substantially reduced if both the components in contact were coated with a material harder than the abrasive. In practice, however, this is difficult to achieve and with typical operating misalignments, edge chipping of the hard and brittle component surfaces is likely to occur.

A more practical solution, therefore, is to harden one component and leave the other one softer. The softer component then embeds some of the abrasive material, which protects it from wear and the overall clearance increase, due to wear, is then determined by the wear rate of the harder component, which although often softer than the abrasive, can provide useful wear resistance.

Figure 2.9 shows the effect on clearance increase of using components of different hardness ratios. It can be seen from this that it is important to avoid the use of components of an equal hardness. In this situation both components embed and displace abrasive material from each other, with mutually destructive results.

In components with an approximately equal contact area on both sides of the moving surfaces, such as pivot pins and bushes, it is generally more convenient to harden the surface of the pin. In the case of scraper blades where the contact areas of the two mating components are widely different, it is generally best to harden the component, such as the blade, which has the smaller contact area.

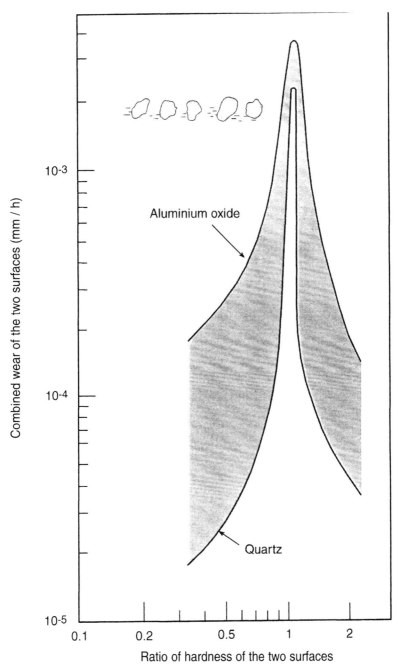

Fig. 2.9 *The wear of two surfaces with abrasive particles trapped between them* *(to see this figure in colour, please refer to colour plate section)*

2.4.3 *Adhesive wear from the rubbing together of smooth surfaces*

This type of wear occurs in machine components which slide over each other and which have a substantial mutual contact area. They range from brakes and clutches, and rubbing bearings operating without lubrication, through piston rings and roller chain link bearings with incomplete lubrication, to high-speed oil-flooded plain bearings, which may experience transient contacts.

In brake and clutch materials the wear can be controlled by arranging for the contact loads to be carried primarily by a fibre or powder structure, supported in a bonding matrix of a dry resin-like material. Similar materials are used for rubbing bearings but the matrix is generally softer and capable of smearing over the load-carrying structure to give reduced friction and low wear under continuous sliding.

For loaded metal-on-metal contacts some lubrication is essential to avoid seizure and wear. The optimum material structure is then one in which there are small dispersed hard areas in the surface to carry the load, supported in a strong but softer matrix, which wears to form recesses around the hard areas, and in which lubricant can be retained. In the case of piston rings and cylinder liners made from cast iron, this is achieved by adding phosphorous, vanadium, or chromium to create localized hard areas of phosphide or metal carbides. In the case of roller chains and other low-speed steel-on-steel bearings the use of hardened carbon steels, with their discontinuous structure, can give adequate performance, but tool steels tend to be better because they contain dispersed carbides, which can create the local load-carrying areas. Similar properties are obtained in materials such as phosphor bronze where the addition of phosphorus creates dispersed hard phases of copper phosphorous compounds.

In fully lubricated high-speed bearings there are thin oil films keeping the components apart but spread over a substantial area. The width of the contact area, relative to the film thickness, makes such bearings very sensitive to misalignment or distortion. This problem is overcome by making the shaft hard, and smooth, and the bearing material soft so that it can conform and thus maintain a uniform load-carrying oil film. The soft bearing material needs to be not more than one-third the hardness of the shaft, and ideally less than one-fifth of the shaft hardness. Such bearing materials need to be strong enough to carry the loading without extrusion but soft enough to be swaged by the shaft for surface adjustment, to allow for misalignment, without risk of seizure. This is achieved in practice by making these materials out of mixtures of hard and soft materials. Two kinds of mixtures are used, one such as the white metals, which have a soft matrix reinforced with a fibrous needle structure of stronger material, the other such as copper lead or lead bronze, which have a strong matrix permeated with soft material that can smear and improve the surface rubbing properties.

This application area of machine components is one in which there are a large number of materials available. The material requirements described above should enable alternative materials to be selected for any applications requiring improved wear performance. Also, since many of the materials are proprietary, comparative wear test results may be available in a number of cases.

A related application area is in press tools which also involve highly loaded metal-to-metal contacts with some degree of lubrication. The material requirements can usually be met by tool steels which have dispersed hard areas in their surfaces. Also, particularly since the contact is intermittent with an opportunity for external lubrication, the use of hard metals or of hard coatings all over the tool surface in the critical areas, is feasible. However, in this case the coating needs to be of adequate thickness, with a progressive transition in hardness between the coating and the substrate. If the coating is too thin or there is a sharp change in hardness at the interface with the substrate, there is a risk that the coating may crack and peel off under impact loading.

2.4.4 *Fretting*

Fretting arises from small-scale movement between surfaces and an important characteristic is that the wear debris is not able to escape easily and therefore continues to take part in the interactions between the surfaces.

This type of wear can occur between pairs of closely contacting machine components which are not intended to move relative to each other, but in fact do so, by small amounts, due to component deflections. Typical examples are: the joint faces and housing bores of assemblies with bolted-on caps, holding dynamically loaded bearings in position; the joints in other structural components subject to superimposed dynamic loads; and contact between the individual wires of dynamically loaded wire ropes or electrical conductor cables. Similar damage can also occur in splined couplings which are subject to small movements in normal operation.

In the case of dynamically loaded fixed joints, when some movement occurs, the high spots of the two surfaces tend to weld together and then break apart, sometimes at a different position, although in many cases the same high spots remain in contact all the time, creating tensile stresses in the surface. Wear debris can also be created, which further aggravate the contact conditions. The process involves local oxidation of the exposed metal and, with ferrous components, when the debris escapes from the area there is often dark reddish brown powdery staining associated with the oxidation of the particles.

There are three undesirable resulting effects which are:

(i)   a tendency for fixed joints to become loose;
(ii)   a tendency for sliding joints, such as splines, to become locked;
(iii)   the creation on the component surfaces of stepped roughness and cracks, which in a dynamically loaded component can lead to fatigue and structural failure.

When the problem occurs there are basically two approaches to solving it:

1. Analyse the strains in a fixed joint under the dynamic loads, and consider whether it is possible to increase the clamping loads between the surfaces and/or increase the component cross-sections in order to reduce the likely deflections.

   This solution may not always be possible because of restrictions on the available space or the allowable mass of a moving component.

2. Improve the conditions at the surfaces of the components so that, although some relative movement may occur, any resulting damage is reduced or ideally eliminated.

   Various techniques have been tried including the use of polymeric or soft metal coatings to provide a separate surface layer in which the movement can take place without significant surface damage. In the case of steel components, various surface treatments can be used such as phosphating, carbo-nitriding, or sulphur addition by various processes. Tuftriding, which is a form of carbo-nitriding, has been particularly effective when used on engine connecting rods.

Fretting problems also occur in wire ropes as used in lifts and cranes. This arises because the individual steel wires in the ropes rub against each other due to the relative deflections which occur when the tension load on the rope changes, or when it bends to enter or leave a pulley or drum. The problem can often be detected by reddish brown oxide powder emerging from the rope (commonly called 'rougeing').

The fretting damage in this case leads to fatigue fracture of individual wires in critical internal positions, and is the determining factor in limiting the life of a rope. Ropes are usually monitored for the presence of broken wires, which can be detected visually or by the use of electromagnetic scanners for checking wire continuity. In critical applications it is also desirable to use rope constructions designed so that wire breaks occur first towards the outside of the rope in order to simplify failure detection.

The problem can be eased by regular lubrication and there should also be potential for surface treating or plating the wires before they are formed into the rope.

2.4.5 Cavitation erosion

Cavitation erosion is a type of wear which occurs on the surfaces of components operating in a disturbed liquid stream. In this situation the flow can create transient zones of very low pressure in which the liquid vaporizes and forms low-pressure bubbles. When these bubbles pass on to regions of higher pressure they tend to collapse implosively, and the liquid then impacts the component surface and subjects it to very high local stress.

The resistance of materials to this type of damage tends to be related to their ultimate resilience, which is the work required to fracture them, or the area under their stress–strain curve up to the fracture point. Figure 2.10 gives some broad guidance on the resistance of various materials to cavitation erosion.

2.4.6 Particle erosion

This is a type of wear which occurs when surfaces are impacted by a continuous stream of hard particles. An example of its intended use is in the sand blasting of surfaces to clean them up by removing surface deposits. It does, however, cause undesirable wear to the inside of pipelines carrying abrasive materials in a fluid medium. Similar problems arise in the flow of crude oil containing sand, where critical control valves can become heavily worn by particle erosion.

The selection of materials to resist particle erosion is critically dependent on the angle of impact of the particles. If the flow is at a

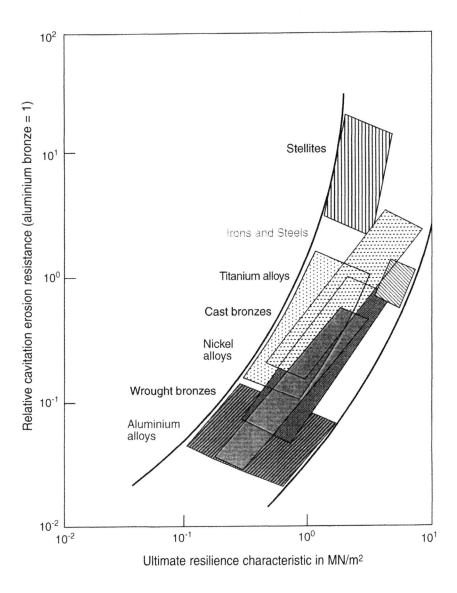

Fig. 2.10 The resistance of materials to cavitation erosion
(to see this figure in colour, please refer to
colour plate section)

shallow angle along the surface, the process is related to abrasive wear and harder surfaces give improved wear resistance. If the flow is nearer to being at right angles to the surface then there is a need for elasticity and resilience in the surface so that the particles are bounced off with the minimum surface damage. For this type of particle flow, materials such as rubber can give the greatest resistance to wear. Since the material choice is sensitive to the angle of approach of the particles, it is generally desirable, in the case of pipelines, to use different lining materials at the bends to that used on straight sections of the pipe.

2.4.7 *Surface fatigue from contact stress*

Surface fatigue is a form of wear which occurs when a surface is subject to a sequence of compressive and tensile stresses. This gives rise to an accumulation of plastic strain which leads to cracking and can then progress to the release of loose lumps of material from the surface. These stress cycles commonly arise from the passage of a concentrated load over the surface such as a loaded ball or roller. In the case of this rolling contact fatigue, improvements are obtained by using materials with the minimum of stress-raising inclusions and with high standards of finish to reduce stress concentration areas in the surface, and to ensure that any lubricant film can keep the surfaces separated.

Since the maximum stresses occur just below the surface it is important with case-hardened surfaces to ensure that the case depth comfortably exceeds the depth where the maximum stress occurs. For rolling bearings and highly rated gears, a minimum material hardness of 58 HRC (Rockwell C) is normally required. An austenitic microstructure is generally desirable for maximum fatigue strength, and this is encouraged in steel by nickel additions in the range 1–3 percent.

2.5 Wear testing of materials

The wear of plant and machinery in industry gives rise to substantial costs. These are sometimes accepted as inevitable, and it would be impossible to eliminate them entirely. However, the costs can generally be reduced, but this requires an understanding and identification of the wear mechanisms involved. This chapter of the guide has given advice on how to do this and on how material changes could help to reduce the wear rates.

This is an area where experience is particularly valuable. However, previous experience may not be adequate to tackle the specific problem encountered. It will then be necessary to obtain a basic knowledge of the wear performance of different materials when subjected to different types of wear. This can be obtained by carrying out comparative and controlled wear tests in laboratories. The various types of wear tests that are available therefore need to be reviewed, and matched to the various practical industrial wear situations. This can then provide guidance on the selection of tests, which are likely to give valid information on the relative wear performance of various materials in practical applications.

The following chapters of this guide review the wear tests that are available in terms of their relevance to practical operating conditions and give guidance on their selection and application.

Chapter 3

WEAR TESTS TO SIMULATE PRACTICAL CONDITIONS

3.1 The wear tests available

A recent survey by the National Physical Laboratory in the UK identified over 400 wear testing standards in use around the world. A review by the American Society of Lubrication Engineers in 1973 identified 300 tests that were in use in various test laboratories. However, many of these tests are slight variations on others, and the true number of tests available is probably less than 100.

In 1997, Neale Consulting Engineers Limited carried out a survey of wear problems in UK industry. The results of this survey have been used to identify the economically significant types of wear. These are the seven types of wear listed in Tables 2.1 and 2.3 of this guide, of which abrasive wear, adhesive wear, and fretting are the most significant. While it may be argued that, in some cases, more than one test method might be justified for a particular type of wear, it is unlikely that much more than ten wear test methods would be required to meet industrial needs for the seven types of wear. There is, therefore, a requirement to select the small number of test methods that are appropriate and, by concentrating on them, to build up a useful background of operating experience.

It is relevant to consider why so many wear tests have been developed when the industrial need is much more restricted. The likely reason is that most of the tests have been developed either by scientific investigators looking very fundamentally at the wear properties of materials, or by particular industries trying to reproduce the operating conditions in some particular application, with very focused test arrangements and procedures.

Surveys have been carried out to compare the results obtained from nominally similar test machines. Pin-on-disc machines have shown variations in the results that are larger than those for many other mechanical properties. The variations are, however, of a similar magnitude to those in the results obtained in other studies of material degradation under aggressive conditions, such as corrosion and creep. These variations, on pin-on-disc machines, tend to be related to the detailed design of the test rigs and associated major variations in the stiffnesses of the test machine components. These have resulted in vibration and load bounce, and transient changes in the intended path of the rubbing interactions. These variations occur particularly with pin-on-disc machines, because their apparent simplicity encourages users to make their own. In contrast, commercially designed specialist standard machines tend to give more consistent results because they are designed and produced by a limited number of specialist suppliers.

The second major cause of these variations is that wear testing is, in any case, more likely to produce variable results than tests for properties such as tensile strength or electrical conductivity, which are carried out under standard conditions on a single specimen. This is because wear tests always involve two specimens between which the wear occurs, and the wear takes place on a micro-scale at the specimen interfaces. The actual contact occurs between a small number (typically four to ten) of asperities and these contact points change from time to time throughout the contact cycle. Also, the exact nature of these contacts inevitably varies from one test to another. There is therefore an inherent variation between tests that will occur however well the machine is designed and operated.

For the best results, both specimens need to be of controlled composition, structure, shape, and surface finish. The surface temperature at the rubbing contacts and the rate of local temperature rise are also very critical factors. These are related to the frictional power intensity and energy pulse input discussed in the following Sub-section 3.2.

These are all very demanding design and control requirements, which may not always be met by test rigs put together in a laboratory, without a full engineering input and analysis. There is therefore a need to study the various wear tests that are available with a comparison of their results and a review of the design of the test machines, to select those few tests that have a good chance of reliably simulating practical machine wear conditions, with a reasonable degree of consistency.

This chapter considers these various issues in order to arrive at a short-list of test methods, which are likely to produce results of relevance to the practical problems of the wear of plant and machinery in industry.

3.2   Important factors in practical simulation

The most realistic wear test on a new material is to test it on the actual machine in service. If the machine is produced in quantity, such as a motor vehicle, it may be possible to do this test on a machine allocated for this purpose, but this takes considerable time and is expensive. A cheaper test is to use the particular section of the machine that is of interest, with its actual components, and run it on some form of test rig. This may allow the test to be accelerated and thus produce results in a shorter time. An even cheaper method is to test the critical component by itself in a rig that attempts to simulate its operating conditions, but this inevitably lacks some realism.

In contrast, a very simple material wear test cannot produce the full operational conditions, but, because of its simplicity, it has the potential to produce results that provide general reference data for a range of similar applications. It is, however, critically important to select a test configuration that is reasonably representative of the practical situation by checking the specimen arrangement and the test conditions to ensure that they represent those in the real machine or machines.

The most important objective in a test is to reproduce the dominant wear mechanism with the surface appearance of the worn test specimens being identical to that of the worn surface of the actual components.

The following factors are particularly important:

1. There will always be two components involved in any wear test, which may be two parts of a machine in mutual contact, or one may be the machine component, with the other being some material with which it makes contact. The materials of these two components in the test need to be very similar to those involved in the actual practical case. Also, the material microstructure and the ratio of the scale of the microstructure to the contact area needs to be the same. If the microstructure is small, so that this ratio is large, then the material can be considered as homogeneous and the test contact area is relatively unimportant. However, with larger microstructures, as in various reinforced plastic bearing materials, the scale size is more critical and the test contact size needs to be similar to that in the application.

2. The component surfaces, at their area of contact, should have a similar shape to those in the service application. If the shape changes with wear, a similar change of shape should occur in the test. Also, the rubbing configuration should be the same. If, for example, the contact is intermittent for at least one of the components involved, then, to simulate this correctly, the actual contact size-to-track length ratio needs to be reproduced in the test.

3. The contact pressure between the two test components should be similar to that which occurs in practice.

4. The pattern of movement and rubbing speeds should be as similar as possible with similar environmental conditions such as temperature, humidity, and possible corrosive interactions.

The pattern of movement combined with the rubbing speeds and contact pressures are particularly significant and need to be compared between the practical case and the test. For example, in practice the surfaces of the two components may be in continuous rubbing contact or only in intermittent contact. In the latter case there are opportunities for surface recovery, oxidation, or cooling, which can have a significant effect on the wear rate.

All wear processes are driven by the dissipation of energy, be they the formation of oxides on the surfaces, the transformation of microstructure, the melting of the surface (which is related to the PV limit* of the material), or surface damage induced by thermal stress. Wear is closely associated with the dissipation of frictional energy in the contact and this is inevitably accompanied by a rise in temperature.

The frictional energy is generated by a combination of load and sliding speed, and its dissipation is influenced by the material properties and the contact area. The wear rate will therefore be expected to be related to these same factors.

---

* For many materials in sliding contact there is a maximum value of the product of applied pressure P and sliding velocity V above which the material fails. This is known as the PV limit.

In most cases the volume of material worn from a surface is broadly proportional to the load and the sliding distance. This is perhaps not surprising, because the true contact area between surfaces depends on the load, and this area multiplied by the distance of sliding relates dimensionally to a volume, such as the volume of material worn away.

The relationships can be expressed as formulae where

W = load
L = distance of sliding
A = area of the wearing surface
t = time
v = velocity of sliding
h = depth of wear
μ = coefficient of friction
k = a constant

Volume wear = kWL

Depth of wear, $h = k\dfrac{WL}{A}$

Depth wear rate = $\dfrac{h}{t}$

$$= k\frac{WL}{At}$$

$$= k\frac{Wv}{A} \tag{3.1}$$

This depth wear rate is the critical factor in practical applications, because it determines the running time to reach a critical amount of wear.

It is also significant in that it relates to the frictional power intensity at the wearing surface. Since

Frictional power input to the surface $= \mu Wv$

Frictional power intensity
(or power input per unit area) $= \mu \dfrac{Wv}{A}$ (3.2)

The similarity of these two formulae [(3.1) and (3.2)] confirms that the rate of material removal is generally directly related to the amount of energy going into a sliding contact.

This is particularly significant when choosing a wear test for a component which has intermittent contact, because it indicates that if a representative depth of wear h_c is to occur during each fractional contact period t_c, then this amount of wear is represented by

$$h_c = k \frac{Wv}{A} \times t_c$$

and there must be an equivalent energy pulse input E_c in the test arrangement of

$$E_c = \mu \frac{Wv}{A} \times t_c$$

In practice, this means that the pattern of contact and non-contact in the test must correspond to that which occurs on the real component, as outlined under point 2 above.

5. The wear process will produce debris and this can interact with the ongoing wear process, either as individual particles or as an agglomeration into larger particles or layers.

It is therefore important to ensure that the wear debris in the test interacts in the same way as in the practical case. The most critical feature is whether the debris remains in the contact in the practical case, or falls away, and this needs to be reproduced in the test.

6. It is always tempting to accelerate a test, to save time, by making the operating conditions more severe. If this is done, it is important to ensure that the wear mechanism does not go through a transition to a more severe type of wear, which would invalidate the results.

For this reason and for general checking for test validity, it is important to examine the worn surfaces from the test in detail to check that the appearance and wear mechanism are the same as those occurring on components from machines that have been in service. An example of this matching is shown in Fig. 3.1(a) and (b). Figure 3.1(a) shows the surface of a die used to extrude concrete components. It is made from a cobalt–tungsten carbide composite and the wear rate varies across the microstructure. Figure 3.1(b) shows the surface of a test specimen from an abrasive wear test that had been adjusted to match the practical conditions, and shows a good reproducibility of the wear process.

Fig. 3.1(a) **Magnified view of the surface of an extrusion die for concrete components**

**Fig 3.1(b) *Magnified view of the surface of a test specimen
reproducing the practical wear conditions***

3.3 The selection of suitable test machines and their important design features

The National Physical Laboratory carried out a survey of users of wear testing machines in the UK in 1997, to determine the types of machines that have been used and any particular problems that have been experienced. They have also conducted a large number of wear tests on a number of the machines, with variations in the operating conditions and the machine design details, to check for sensitivities and likely errors in results. The combined experience from this is included in this section of the guide, which recommends appropriate test methods for the various types of wear.

3.3.1 Abrasive wear by hard, granular materials

The test arrangements that have been used for this type of wear are shown in Fig. 3.2.

A critical feature of abrasive wear tests using hard, granular materials is that, if the material is used for more than one contact with the test specimen, it can become modified and give unrepresentative results. If a loose, sharp abrasive is being used it may become partially blunted or may be broken up into smaller sized particles. The overall effect of this will be to give a reduced wear rate on the specimen compared with that which would have occurred if new, sharp, abrasive material had been continuously fed to the specimen. This is a problem with the use of wear tests such as that shown in Fig. 3.2(c). An alternative simple test is to use the abrasive bonded to a backing material, similar to abrasive papers, but the problem then is that the abrasive becomes blocked with wear debris as well as the previously described risk of blunting. This situation invalidates tests such as that shown in Fig. 3.2(a). The test shown in Fig. 3.2(b), which uses a spiral track, overcomes this problem but the length of available wear track then provides a limit to the applicability of this test.

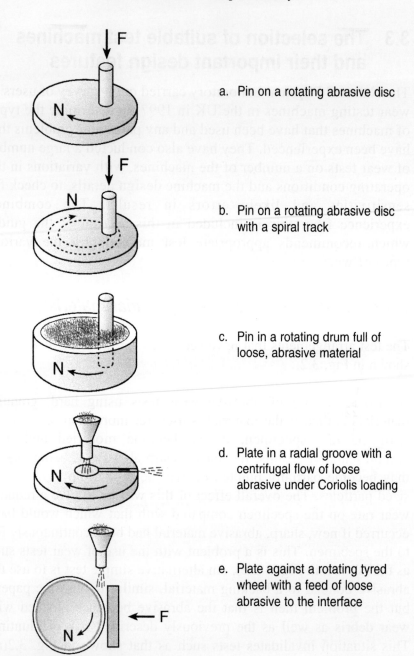

a. Pin on a rotating abrasive disc

b. Pin on a rotating abrasive disc with a spiral track

c. Pin in a rotating drum full of loose, abrasive material

d. Plate in a radial groove with a centrifugal flow of loose abrasive under Coriolis loading

e. Plate against a rotating tyred wheel with a feed of loose abrasive to the interface

N indicates rotation
F indicates externally applied force

Fig. 3.2 *The configuration of various test rigs for abrasive wear from contact with granular material*

Also, in practical applications involving more rounded grits, the grit will fracture into sharper particles if the load is above a critical value. This again requires a continuous feed of fresh grit and the careful selection of the test load.

Tests which use a continuous feed of new, granular abrasive therefore have a wider range of application. The arrangement shown in Fig. 3.2(e) can be operated with a steady and controlled known load, while the arrangement shown in Fig. 3.2(d) creates a load from the circumferential acceleration that needs to be given to the particles as they move out to rotate at an increased radius. The system in Fig 3.2(d) is a less controlled process than that in Fig. 3.2(e), which is therefore a more accepted and reliable method for abrasive wear testing. It also has the merit that the rotating wheel can be tyred with various materials. Rubber can be used to simulate the properties of abrasive materials which in practice have a semi-elastic backing, such as bulk particulate materials or extrudable materials with a clay type of matrix.

This test has been standardized as ASTM G65 using a rubber-tyred wheel 228 mm in diameter running at 200 r/min, fed with a sand abrasive about 200 μm in size at a rate of 300–400 g/min. The test is run for a set number of wheel revolutions, typically in the range 1000–6000, with a test load of between 45 and 130 N. Since the wear zone is of a variable profile, the amount of wear is measured by weight loss or volume.

A critical feature of this test machine is the design of the sand feed nozzle because the feed must be uniform across the wheel width and at a steady rate. The contact load also needs to be carefully controlled.

Another problem is that the abrasive particles can slide across the test surface with a cutting action, or can roll across resulting in material displacement, rather than cutting. It is, therefore, important to examine the specimen surfaces after the test to ensure that the mechanism is consistent across the test surface and similar to that on the actual component in service.

Another possible source of error, when testing the wear resistance of composite materials consisting of hard particles in a tough matrix, is that the wear performance depends on the relationship between the hard particle spacing in the test material matrix and the abrasive particle size. This confirms the importance of checking that the particle size, shape, and hardness of the test abrasive is representative of that which will be met in the practical application. The ASTM G65 test, as shown in Fig. 3.2(e), can also be adapted for wet testing, and with liquids of various corrosive properties, so that interaction between abrasive wear and corrosion can be checked.

A similar general test configuration to Fig. 3.2(e) can also be used to simulate wear by abrasive slurries, such as can occur on the inside surface of pipes, particularly when the flow direction changes. In this case the wheel is rotated in the opposite direction to lift the abrasive up from the bottom of a test tank in which the whole system is immersed in a wet abrasive slurry mixture. This has the disadvantage that some of the abrasive is recirculated and can become blunt. However, as in the ASTM G65 test, the liquid can be selected to be corrosive, if this is typical of the practical arrangement. This can therefore give the corresponding increased wear rates that would then be expected. Wear tests in this category are ASTM G105 which uses a rubber wheel and ASTM B611 which uses a steel wheel.

As always, the test specimen needs to be examined to check that the surface appearance is similar to that of the actual component for which the wear is being simulated. If unexpected differences do occur, this can be due to the abrasive forming a bed on the surface of the wheel, as distinct from flowing through as a series of individual particles.

A type of abrasive wear that is difficult to simulate is that which involves heavy impact loading, such as occurs in materials handling and the crushing of rock-like materials. In these cases, useful results can often be obtained by inserting test tiles, of alternative materials, in the operating surfaces of the actual machines.

3.3.2 *Abrasive wear by particles trapped between surfaces*

The test arrangements that have been used for this type of wear are shown in Fig. 3.3.

An important feature of this type of wear is that, as shown in Fig. 2.9, the two surfaces need to be of different hardnesses in order to obtain lower total wear rates of the two components. The mechanism then is that the abrasive embeds in the surface of the softer component and tends to protect it from wear, with the wear rate being determined, primarily, by the properties of the harder component. In any test, therefore, the softer surface needs some degree of free exposure to the abrasive so that it can embed the abrasive when rubbed by the harder component. In tests such as those shown in Fig. 3.3(a) and (b), this means that the harder of the two components needs to be represented by the block or pins and the softer component by the plate or disc. This also corresponds to the actual configuration in machines, such as plaster mixers, where the smaller area components, such as scraper blades, are made of hard material to resist wear from their contact with other components, such as discs, of a much greater surface area.

The test shown in Fig. 3.3(a), with its reciprocating motion, is particularly appropriate for wear situations involving oscillating movement such as linkage pins. For such a test it is desirable to use a block with a slightly curved contact surface to simulate the contact entry shape between a pin and its hole with a corresponding diametral clearance.

The test shown in Fig. 3.3(b), with its continuous rotation, is more appropriate for the simulation of higher speed, continuously moving contacts such as the scraper blades of mixers. In such applications, the deflection of the blade under the influence of friction will be an important practical feature in its operation – this can be simulated in the test, by the use of pins of an appropriate lateral stiffness.

It is also possible to use tests such as ASTM G65, as discussed in the previous section, to simulate this type of wear using metal wheels and tests blocks of different hardness.

The tests shown in Fig. 3.3(c) and (d) have some features in common with practical situations where hard particles become embedded in one surface and then wear another. Typically, however, the test ball is one as used in a ball bearing, in the interest of specimen standardization. The ball hardness will therefore be at least 60 Rockwell C or 750 Vickers and since, for simulation, this abrasive-carrying surface needs to be the softer of the two, the results from this test are only likely to be relevant directly to practical conditions, in which one component is very hard, such as a ceramic coating. Alternatively, softer metal balls can be used, but their accurate manufacture requires special skills.

However, the real merit of this particular test is that it provides a wearing 'indentor' of constant geometry. This produces a spherical wear crater, which in the case of thin surface coatings exposes the interface and substrate in a controlled manner.

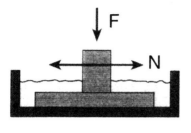

a. Block on a plate reciprocating in a slurry

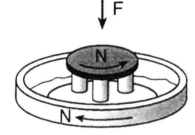

b. Three pins and a disc in relative rotation in a slurry

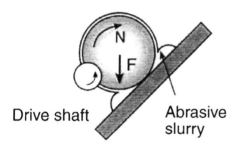

Drive shaft   Abrasive slurry

c. Rotating polished steel ball rubbing on a test surface, in the presence of an abrasive paste, to produce a circular crater

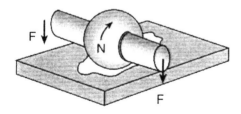

d. Rotating end clamped ball rubbing on a test surface in the presence of an abrasive paste to produce a circular crater

N indicates movement
F indicates externally applied force

Fig. 3.3 The configuration of various test rigs for abrasive wear by particles trapped between surfaces

This test configuration has, therefore, been found to be particularly relevant as a test for surface coatings as such, without necessarily simulating exactly any practical configuration. When testing surface coatings it is important to check, for the test loads and conditions, the depth at which maximum Hertzian contact stresses will occur. If this maximum stress occurs near the coating–substrate interface it can give rise to bonding problems. Ideally, the maximum contact stress should be at the same depth on the test as it is in practice, if this is known.

The drive arrangement shown in Fig. 3.3(c), while simple, does have the disadvantage that the driving torque on the ball affects the contact load on the test surface, which therefore varies with the friction between the ball and the surface. An alternative arrangement as shown in Fig. 3.3(d) with the ball clamped end-wise between two co-axial shafts, externally loaded on to the test surface, has been used to overcome this problem. The ball size and material then become independent of the load and this results in a more versatile testing machine.

3.3.3 *Adhesive wear between two surfaces*

The test arrangements that have been used for this type of wear are shown in Fig. 3.4.

Adhesive wear is the most common mechanism for the wear of machine components such as bearings, piston rings, and roller chains. With these components there is generally a lubricant present and many practical tests are concerned with the testing of lubricants to prevent wear occurring. This guide is, however, concerned with wear testing, which involves tests on materials to ensure that in the event of lubricant breakdown, the wear will be minimized. As in the abrasive wear tests an objective of the test will be to produce on the rig similar surface damage to that which occurs after a failure in service. This may require the presence of very small amounts of the lubricant used in service to achieve the same contact chemistry.

There are also components such as brakes and clutches and dry rubbing bearings, which operate without lubrication, and for which adhesive wear tests are directly relevant to their normal in-service performance.

With disc brakes the actual configuration of brake pads and discs is sufficiently simple for direct use for the wear testing of materials. A critical testing decision will then be the operating temperature at which the test is carried out, because there will be substantial heat generation from friction in any continuously running test. If the testing is carried out on a simplified rig, it will be important to maintain the same ratio of pad contact length to total disc track length, so that the material is subject to similar thermal cycles to those occurring in the actual brake.

In the case of dry rubbing journal bearings and thrust washers, the actual configuration is again relatively simple and arrangements such as those shown in Fig. 3.4(d) and (e) are commonly used for component material wear tests. The arrangement shown in Fig. 3.4(d) tends to give more consistent results than that shown in Fig. 3.4(e) because in the latter arrangement the contact area between the components varies substantially during a test. This is because, in practice, there has to be some diametral clearance between the specimens to allow for assembly, so that the rubbing contact changes from an initial nominal line contact to an area contact as the components bed in to each other. This does, however, represent the actual configuration for a practical journal bearing, so it is not an irrelevant test. The constant contact geometry arrangement represented by the arrangement shown in Fig. 3.4(d), however, does enable the basic material wear factors to be evaluated more directly. In the journal bearing test it will be important to have a shaft and housing arrangement of a similar thermal dissipation capacity to that in a typical application. This will become more critical if the operating conditions are made more severe to the point where surface melting of the bearing material may begin to become a factor.

Similarly, the arrangement shown in Fig. 3.4(f) can give an effective simulation of a piston ring and cylinder liner in rubbing contact.

However, the test specimens are of a relatively complex geometry, which is a disadvantage if new materials are to be tested. Tests on existing materials can conveniently use test specimens cut from the actual components. However, if an actual piston ring is used with a matching liner it will be found that the free ring curvature does not match that of the liner, when the ring is not radially compressed as in the piston. A practical solution to this is to use test pieces cut from the ring next to its gap and discard the rest of the ring. If new materials are to be tested, a specimen with a curved edge can be cut from a sheet of the new material, but the bending and twisting stiffnesses are likely to be different from a ring sample and this can affect the results.

The simpler test configurations shown in the arrangements represented in Fig. 3.4(a), (b), and (c) do have advantages in terms of their basic general reproducibility. When using them, however, it is essential to reproduce the practical contact conditions as closely as possible. This also includes the same motions as those experienced in service, which can critically affect the results in some applications. For example, in wear tests on materials for replacement hip joints, a simple uni-directional oscillating movement can give much less wear than the practical situation in which the oscillating movement is multi-directional. This arises from the particular wear properties of the plastic materials involved, in which the long chain molecules become aligned with the direction of motion, and if this is uni-directional, it increases the wear resistance.

A pin-on-disc test as shown in the arrangement depicted by Fig. 3.4(a) has been used as a very simplified test for piston ring and cylinder materials with the pin representing the ring and the disc representing the cylinder. The pin has a continuous rubbing contact and the disc an intermittent contact, and the pin diameter and its rubbing radius should be matched to the practical configuration. The contact surface of the pin can also be spherically shaped to represent the running surface of a piston ring, which is typically similar to a barrel with a large radius of curvature. However, the true surface finish on the cylinder liner cannot be reproduced easily.

A problem with this arrangement [Fig. 3.4(a)] is that the running temperature tends to be controlled by the frictional heating and this may not match the practical values combined with typical contact loads and speeds. This can be overcome by the use of the arrangement shown in Fig. 3.4(c), which can attempt to simulate the conditions for a piston ring around top dead centre, where the wear usually occurs. The reciprocating velocities will be lower than the rubbing speeds used in the arrangement of Fig. 3.4(a) and the temperature can be externally controlled.

The arrangement shown in Fig. 3.4(b) provides a simplified configuration of the arrangement represented by Fig. 3.4(e), in that a test material only needs to be available in the form of a block and not a complete bearing bush. The block can also be pre-shaped to give a consistent bedding area through the test. However, in practice, there are invariably problems with misalignment and the block mounting needs to have some provision for lateral self-alignment. Running in will also be needed before a valid test for wear rate can be started.

Simplified tests such as the arrangements shown in Fig. 3.4(a), (b), and (c), to simulate the adhesive wear of machine components, need to operate under conditions that simulate those in the practical situation. This means the same materials and surface finishes, contact pressures, and sliding speeds and motions, temperatures, and contact patterns in terms of continuous or intermittent contact for each of the interacting components represented. In addition, however, the potential wear debris interaction should be the same as in the practical case. This means that, if the wear debris remains around the contact area in practice, it should also do so in the test, and the arrangements shown in Fig. 3.4(a) and (c) will generally allow this. If in practice the wear debris is free to fall away, then the arrangement shown in Fig. 3.4(b), or alternatively those in (a) and (e) turned through 90 degrees to make the plane of the disc or plate vertical, are likely to reproduce the practical conditions with greater accuracy. It also needs to be remembered that while these tests are to measure the adhesive wear resistance of materials, in practice the surfaces may be lubricated. It may therefore be desirable to carry out repeat tests with lubricants present on the surfaces.

It is also critically important with these simplified tests that the whole test machine is designed with a high level of engineering understanding. For example, the friction forces can be high and potentially variable, which can give rise to vibration as well as resulting wear track variation and friction measurement errors. Poorly designed machines can give wear results that are a factor of ten different from others, because of these effects. The support stiffnesses of the two sides of the rubbing contact are the most critical features. For example, on a pin-on-disc test machine, as shown in Fig. 3.4(a), the shaft and bearings supporting the disc must be of adequate diameter and the arm supporting the pin must have its stiffness carefully considered. If the friction forces on the pin from the moving disc are carried as torsional loads in the arm, the support will tend to be more flexible than if they are carried in bending. This is determined by the configuration of the arm relative to the contact point of the pin on the disc. Another critical design feature is the height of the pin support arm axis above the pin and disc contact point. Ideally this needs to be kept to a minimum, to reduce the input torque arm length. Any friction force measuring system also needs to be very stiff so that support rigidity is not lost.

If in a particular test it is necessary to allow for some flexibility in the system, to represent a particular practical case, this can easily be achieved by a reduced pin cross-section at some point in its support length. This is a controlled feature, as distinct from the built-in flexibility and poor specimen location resulting from inadequate test machine design.

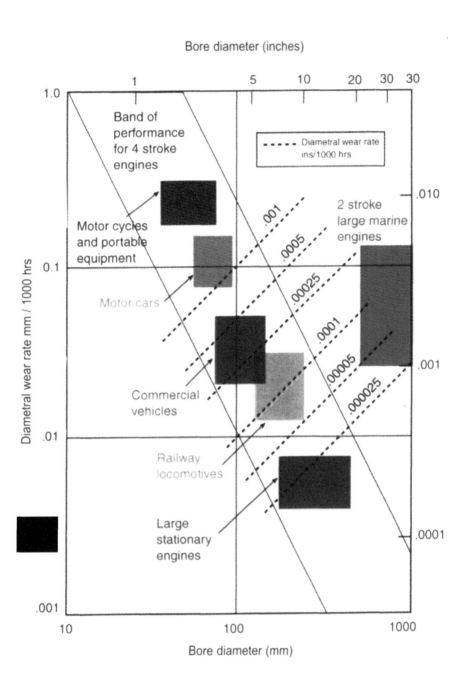

Fig. 2.8 The typical wear performance of the cylinders of internal combustion engines
(From Chapter 2, Industrial Wear Problems, page 19)

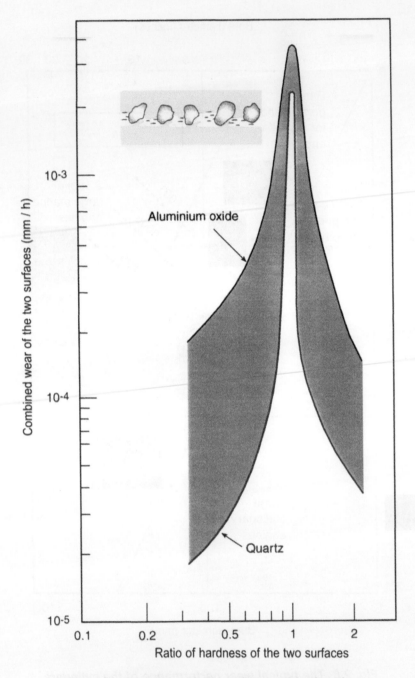

Fig. 2.9 The wear of two surfaces with abrasive particles trapped between them
(From Chapter 2, Industrial Wear Problems, page 23)

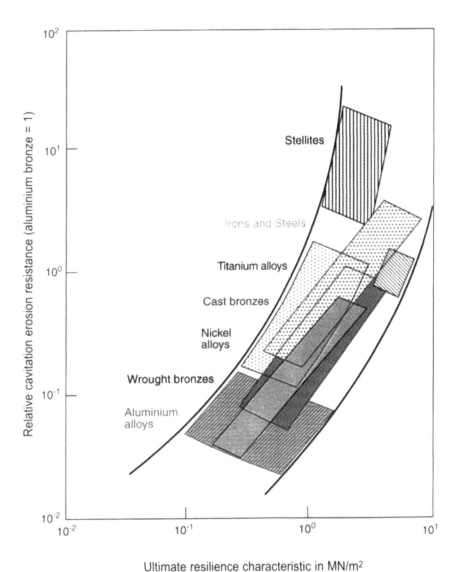

Fig. 2.10 The resistance of materials to cavitation erosion
(From Chapter 2, Industrial Wear Problems, page 29)

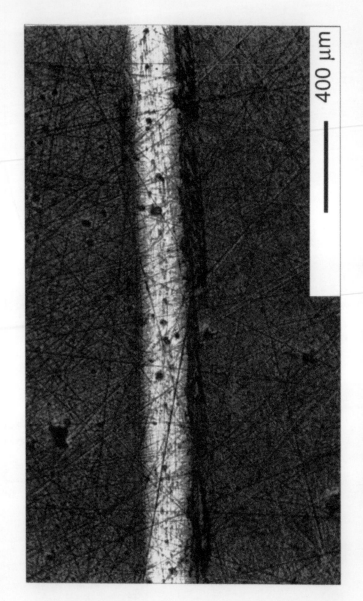

400 μm

(a)   *Light wear showing build up of titanium oxide at boundaries of wear track*

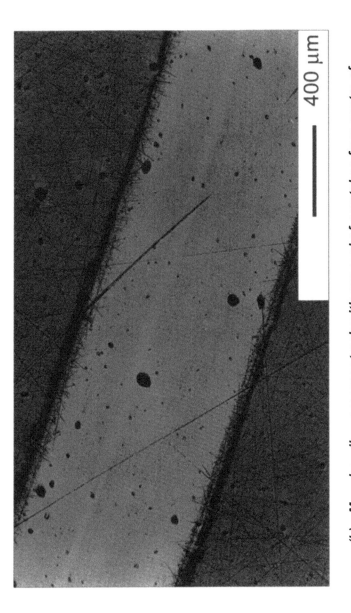

400 μm

(b) *More heavily worn wear track with removal of scratches from centre of wear track by wear process*

Fig. 5.2 High-magnification images showing surface features. Wear track on titanium nitride-coated samples worn in pin-on-disc configuration (from Chapter 5, Examination of the Worn Surfaces, pages 78 and 79)

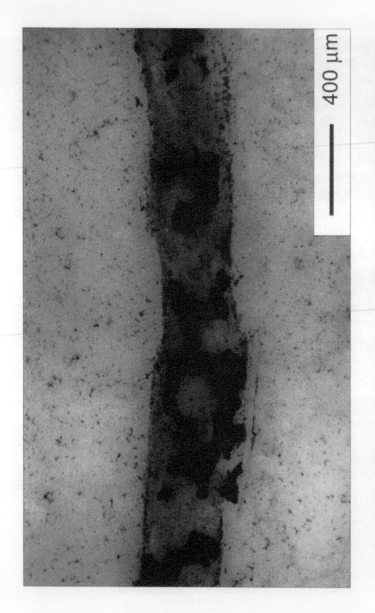

400 μm

(a) Imaged with near-crossed polarizer illumination and emphasizing the iron oxide reaction layers on the surface of the wear track

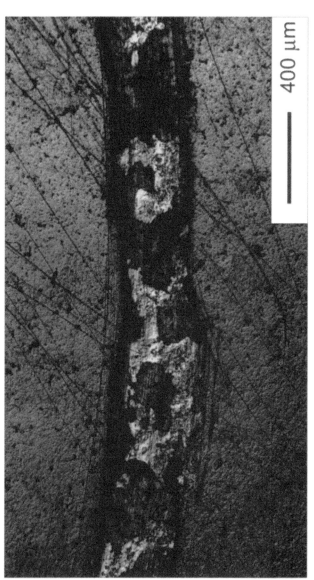

─── 400 μm

(b) *The same area imaged with differential interference contrast emphasizing the transferred layers of steel on the wear track*

Fig. 5.3 *The effect of different lighting arrangements on the images. Iron- rich transfer films on alumina sample formed during wear of steel pin against alumina sample*
(From Chapter 5, Examination of the Worn Surfaces, pages 80 and 81)

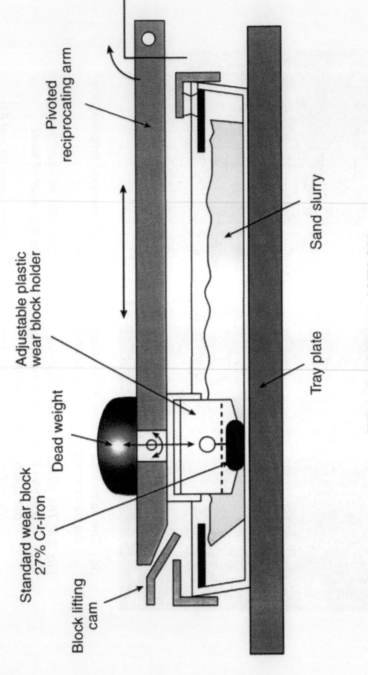

Miller abrasivity test, ASTM G75
(From Appendix A, page 104)

Block lifting cam

Standard wear block 27% Cr-iron

Dead weight

Adjustable plastic wear block holder

Pivoted reciprocating arm

Sand slurry

Tray plate

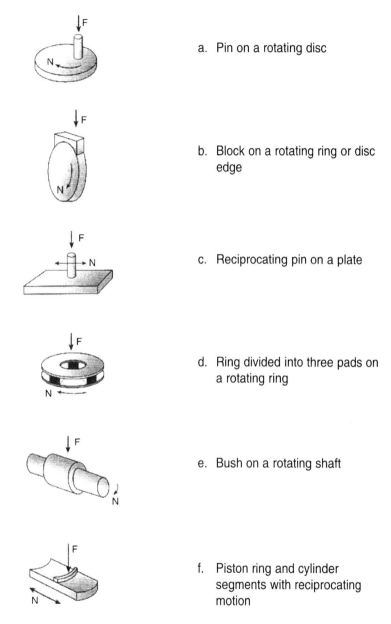

a. Pin on a rotating disc

b. Block on a rotating ring or disc edge

c. Reciprocating pin on a plate

d. Ring divided into three pads on a rotating ring

e. Bush on a rotating shaft

f. Piston ring and cylinder segments with reciprocating motion

N indicates movement
F indicates externally applied force

Fig. 3.4 The configuration of various test rigs for adhesive wear between two surfaces

3.3.4 Fretting

The test arrangements that have been used for testing materials and surface treatments for their resistance to fretting are shown in Fig. 3.5.

Fretting occurs when there are small oscillatory movements between heavily loaded metal surfaces. These are typically the joint faces of components clamped together or fitted with an interference fit. Similar effects occur between the individual wires of wire ropes and electrical conductor cables.

A simulation test needs, therefore, to rub components together with a reciprocating motion of small amplitude, typically less than 250 μm, while a high mutual contact pressure is applied between them. In most practical applications, other than wire ropes, the full contact length between the components is typically at least one hundred times this amplitude of oscillating movement, but to use test

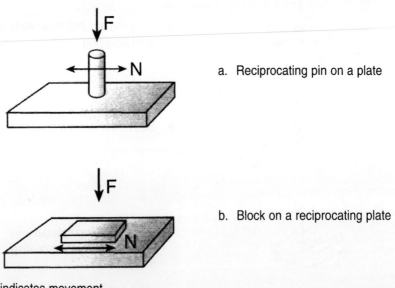

a. Reciprocating pin on a plate

b. Block on a reciprocating plate

N indicates movement
F indicates externally applied force

Fig. 3.5 The configuration of test rigs for fretting between two surfaces

specimens of this size would require excessively high loads for a typical test machine. In practice, the damage patterns produced by fretting are of a random nature, but a typical scale length of observed damage patterns on surfaces is probably of the order of 2.5 mm. The contact zone of test specimens should therefore be of at least this size, and the amplitude of the reciprocating motion may need adjustment to produce representative damage patterns.

Either of the arrangements shown in Fig. 3.5(a) or (b) can be used, but the machine design needs to be rigid to ensure that the intended small amplitude of the movements is actually applied between the test specimens. As with other wear processes, fretting is affected by temperature and it is therefore important to carry out the test at a similar temperature to that occurring in service. The frictional power intensity [see formula (3.2) in Section 3.2] that is estimated to be occurring at the contact, can also provide a guide to the testing conditions.

The particular result of interest in fretting tests is not usually so much any actual wear rate, but is more related to the surface finish changes and surface cracking which the fretting tends to produce. This is particularly significant because in dynamically loaded components, it is these features which give rise to subsequent component structural failure by fatigue.

3.3.5 Cavitation erosion

A test arrangement that can be used for cavitation erosion tests is shown in Fig. 3.6. In this arrangement the cavitation is created cyclically by the use of a high-frequency oscillatory drive, commonly using magneto-striction or piezo-electric effects. This causes the test surfaces to approach and recede from each other at high speed. During recession the fluid between the surfaces cavitates, and on approach the cavitation collapses and erodes the surfaces. To achieve these effects hydrodynamically, the specimen dimensions at the contacting surfaces typically need to be at least 10 mm in diameter and flat.

Another possible cavitation erosion test is that based on ASTM G134 which uses a cavitating liquid jet. The damage patterns produced by such a test tend to be on a smaller scale than those arising in practical applications. They tend to appear more like a coarse matt finish rather than the holes of 1 mm or more diameter generally observed in practice. In many cases, this difference may be unimportant but errors could arise in tests related to components with larger scale compositional or grain size variations.

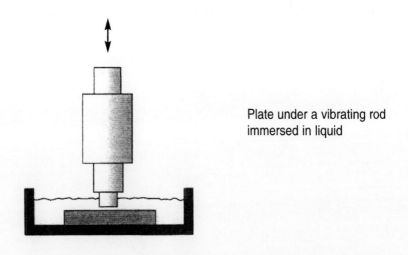

Plate under a vibrating rod
immersed in liquid

Fig. 3.6 The configuration of a test rig for cavitation erosion

3.3.6 Particle erosion

Test arrangements used for this type of wear are shown in Fig. 3.7. They are both very similar, but the arrangement shown in part (a) is used for liquid jet erosion, and (b) for gas-driven jet erosion.

The important test requirements are that the abrasive must be representative of that occurring in the actual application in terms of its particle size, shape, and hardness. It must only be used once, to avoid the risk of blunting or particle fracture into smaller grit sizes. The velocity of the particles (not the speed of the gas or liquid) is critical and needs to be representative of the practical case.

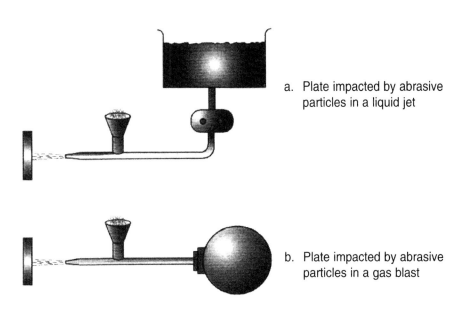

a. Plate impacted by abrasive particles in a liquid jet

b. Plate impacted by abrasive particles in a gas blast

Fig. 3.7 The configuration of test rigs for particle erosion

The grit feed system is a critical feature of the machine design because it needs to provide a steady uniform flow of the grit into the fluid flow. Some standard tests use nozzle diameters of the order of 1.5 mm, which require the abrasive particles to be fed in under pressure. Better results are often achieved with larger diameter, long nozzles in which any nozzle wear is less significant.

Most tests are conducted with the target test material surface at right angles to the jet direction, but this can be varied to attempt to simulate a practical configuration or to investigate the effect of approach angles. With a specimen at right angles to the jet, the wear crater will tend to have inclined sides around its edges largely associated with the escape path of the particles, and in any test any change in contact angle arising from wear needs to be considered when interpreting the results. Because of these variables it is usually necessary to carry out a number of tests to ensure that the conditions in the practical application are properly represented.

3.3.7 Surface fatigue

The various test arrangements used for surface fatigue tests are shown in Fig. 3.8. These tend to be tests on actual machine components as in the case of Fig. 3.8(c) and (d). The tests shown in Fig. 3.8(a) and (b) are an approach to simplified test arrangements and have been used for the accelerated testing of rolling bearing materials for their fatigue strength properties.

These tests tend to be used by the manufacturers of rolling bearings and plain bearings, and some user industries. They are not in general use in testing laboratories.

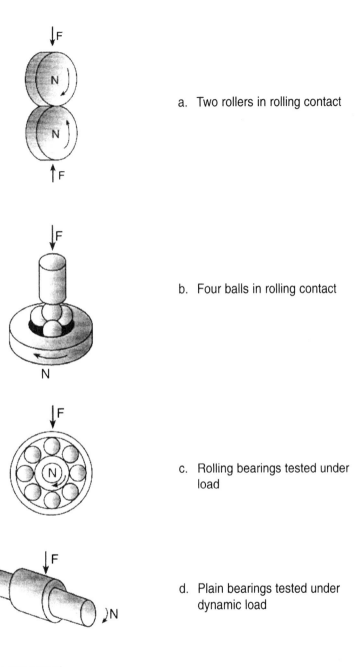

a. Two rollers in rolling contact

b. Four balls in rolling contact

c. Rolling bearings tested under load

d. Plain bearings tested under dynamic load

N indicates movement
F indicates externally applied force

Fig. 3.8 The configuration of test rigs for surface fatigue

3.4 General procedure in tests to ensure simulation

The specific object of wear tests is to give guidance on the likely performance of alternative materials in service, so that the test must be representative of those service conditions.

The first step, therefore, is to identify the type of wear that is occurring on the actual component by careful examination of its surface and to compare this with the descriptions of the nature of the wear damage given earlier in this guide, together with the photographs in Figs 2.1 to 2.7. Once this has been determined, the appropriate wear test can be selected as indicated in the previous section of this guide.

A test programme then needs to be devised to produce the same contact conditions of friction power intensity and energy pulse input, together with the conditions of continuous or intermittent contact, that apply in the practical application.

A number of preliminary tests then need to be conducted, to reproduce the same wear patterns, using materials that are identical to those used in current service, and the specimens examined after each test for comparison with the actual specimen from operating service. Useful guidance can also be obtained by checking the wear debris that has been created and comparing it in size, shape, and composition to any wear debris samples that may be available from operating service. It is essential that the same specimen surface appearance and general condition is reproduced in the test, and it may be necessary to repeat the preliminary tests several times with different operating conditions until this is achieved. This situation can arise because it is often difficult to identify fully what the real contact conditions actually are, and then transfer them to the test machine. However, once the actual wear has been reproduced, the test conditions required to produce this can give an insight into the actual nature and causes of the practical problem.

Once a representative test procedure has been established, tests can then start on alternative replacement materials, with the object of identifying any which offer the possibility of improved performance in service. When a material that is noticeably better has been identified, it can be useful to test it as soon as convenient on the actual component in practical service to see whether a correspondingly better result is achieved. If it is, then it can be concluded that the test and its conditions are reasonably representative and a more comprehensive test programme can then be started.

In practice, it is unusual to be able to obtain a perfect reproduction of the service situation in a laboratory test and some degree of imperfection usually has to be accepted. Inevitably, a fair amount of guided and informed judgement is needed in interpreting the results, and any experience with the use of the new materials in other practical applications needs to be fed into the overall investigation.

Chapter 4

A GUIDE TO WEAR TEST
SELECTION

This chapter is intended to provide assistance with the rapid selection of a suitable wear test method to match a variety of practical industrial situations. Most practical situations are included in the selection tables given on the following pages and, once a possible wear test method has been identified, further information on its features and method of use can be found in earlier chapters of this guide, and in the appendices. In all cases, the test conditions will need to be arranged to simulate the actual practical conditions, and then adjusted to obtain the same surface conditions on the specimen as are observed in practice.

Wear test selection tables

Type of wear	Actual wear process	Typical practical examples	Operating conditions
Abrasive wear	The wear of machine or plant components from moving contact with hard granular materials	Rotors of powder mixers Extrusion dies for brick making Material handling chutes	Abrasive grits up to 0.5 mm in size and sliding speeds up to 3 m/s
		Teeth of digger buckets	Abrasive stones and rocks with sliding speeds up to 1 m/s
	The wear of machine components in mutual moving contact but with abrasive material getting into the contact	Pivot pins in construction machinery	Abrasive grits up to 0.25 mm in size. Slow oscillating motion and high contact loads
		Scraper blades in mixing machines	Abrasive grits up to 0.25 mm in size. Sliding speeds up to 3 m/s

Appearance of worn surfaces	Test method	Remarks
Fine scratches all over the surface	Rubber wheel abrasion test ASTM G65 No. 2 test in Appendix A	If in service the abrasive is very rigidly supported, a steel wheel abrasion test may give better simulation
Scratches and gouges on the surface	A rubber wheel or steel wheel abrasion test may give some guidance using similar abrasive in a smaller size than that involved in practice. Spiral track pin-on-abrasive-disc tests can also be applicable	A better test may be to fit actual teeth of different materials for simultanous comparative tests in service
Scratching and scoring in the direction of motion	Block on a plate reciprocating in a slurry ASTM G75 No. 5 test in Appendix A	The pin should be harder than the hole or bush in which it operates
Scratching and scoring in the direction of motion	Steel wheel wet abrasive slurry test ASTM G105 No. 4 test in Appendix A or Struers three-pin-on-disc test as Fig. 3.3(b)	Simultaneous comparative tests on different blade materials in service may also be possible

Wear test selection tables (continued)

Type of wear	Actual wear process	Typical practical examples	Operating conditions
Adhesive wear	Low-friction non-metallic materials in rubbing contact with metals	Rubbing bearings Electric motor brushes	Up to 50 MN/m² and up to 10 m/s but not simultaneously. PV values up to 1.5 MN/m² x m/s
	High-friction non-metallic materials in rubbing contact with metals	Brakes Clutches	Up to 30 m/s and up to 2 MN/m²
	Metal components in rubbing contact with each other due to inadequate lubricant film thickness	Bearing surfaces in machines	Up to 40 MN/m² and 0.5 m/s
		Valve spindles and stems	Up to 2 m/s reciprocation and 1 MN/m²
		Piston rings and cylinder liners	Up to 7 MN/m² and 40 m/s but not simultaneously

Appearance of worn surfaces	Test method	Remarks
Polished with some fine scoring	Bearing bush test Fig. 3.4(e) Block on ring test Fig. 3.4(b) Thrust washer test No. 9 in Appendix A	The test may require cooling to reproduce representative temperatures. High temperatures will give incorrect results
Polished with some scoring	Thrust washer test No. 9 in Appendix A Block on ring test Fig. 3.4(b)	Tests in actual brakes and clutches with appropriate temperature control are also possible
Scuffing and smearing	Pin-on-disc test No. 7 in Appendix A Reciprocating test No. 8 in Appendix A	Representative service conditions are somewhat indeterminate. Test must reproduce service surface damage
Scuffing and smearing	Reciprocating test No. 8 in Appendix A Pin-on-disc test No. 7 in Appendix A	Ditto
Scuffing	Reciprocating test No. 8 in Appendix A Pin-on-disc test No. 7 in Appendix A Ring and liner segments reciprocated as in Fig. 3.4(f)	Ditto

Wear test selection tables (continued)

Type of wear	Actual wear process	Typical practical examples	Operating conditions
Fretting	Small movements between clamped surfaces (movement small in relation to contact size)	Connecting rod joint faces and bores in internal combustion engines Contact between wires in wire ropes	Movements up to 250 μm Contact pressures up to 200 MN/m²
Cavitation erosion	Collapse of vapour bubbles in a fluid stream impacts the surface to create part-spherical cavities	The blades of ship's propellers Pump rotors Control valves for liquid flow	Rapid fluid pressure changes associated with turbulence and flow breakaway
Particle erosion	The wear of surfaces by the passage of fine sharp abrasive materials in a high-speed fluid flow	High-pressure control valves for crude oil containing sand Helicopter rotors in desert operation	Up to 50 m/s flow with suspended abrasive particles

Appearance of worn surfaces	Test method	Remarks
Roughened with small pieces of transferred material stuck to the surface. Possibly some surface cracks	Fretting test No. 10 in Appendix A Reciprocating test No. 8 in Appendix A	The nature of the surface damage rather than the wear rate is what the test should reproduce
Pitting with cavities usually made up from a series of part- spherical eroded areas	High-frequency reciprocating contact in fluid No. 11 in Appendix A	The surface damage produced in the test will tend to be on a smaller scale than that occurring in practice
Extensive fine scratching/pitting leading to extensive material removal	Liquid and gas blast erosion tests No. 12 and No. 13 in Appendix A	The angle of flow impact occurring in practice needs to be reproduced in the test to get a true simulation

Chapter 5

A GUIDE TO THE EXAMINATION OF THE WORN SURFACES

As soon as any wear test has been completed, the results need to be validated by examining the worn specimens, to ensure that the wear mechanism in the test is the same as that occurring in service. The specimens from the test and the worn components from service both, therefore, need to be examined and compared. This chapter gives detailed guidance on the procedures that should be followed when doing this.

5.1 Sample preparation

Samples obtained from laboratory wear tests have a clearly defined history: the worn area on the sample is often very clearly defined, and little special preparation is needed. However, if there is a large amount of loose debris and other material on the surface, this may need to be removed to reveal the underlying structure (although debris should always be retained for examination).

The preparation of worn samples and components, which have been extracted from field trials or from real application environments, is much more complex. The parts are often covered with oils, grease, or other material from the working environment, which needs to be removed by careful cleaning to reveal the worn surfaces. The material removed from the surface should be retained, as it may be needed for future reference. Information from relevant samples can be lost if they are inadequately stored. Corrosion and other subsequent damage can then destroy the original worn surface. It is always important to obtain as much information as possible about the situation of the worn part or component, so that the best understanding of the mechanism of damage to the sample or component can be obtained. The worn components should also be compared with similar new components, so that the actual changes in dimensions and appearance due to wear can be recorded.

Cleaning is normally carried out after removal of any loose material or wear debris by washing with organic solvents, such as alcohols or acetone. For small samples, ultrasonic agitation can be beneficial.

In some cases, particularly for samples from field trials or normal service, the actual worn component cannot be easily obtained. This can be either because the components are very large and sectioning them can be expensive and time consuming, or because the parts concerned cannot be removed from service. In these cases, replicas can be formed with solvent softenable plastic films or by applying cold-setting moulding compounds to the worn surfaces of the components (after suitable cleaning). These can then be examined by a range of the techniques described in this report, although clearly compositional and mechanical property information cannot be acquired. It is also essential to take photographs on site and to record all the significant operating conditions.

5.2 Surface appearance

5.2.1 Visual examination

It is always important to examine worn samples and components by eye before using any other technique. This visual examination gives information about the whole component that other, higher resolution techniques can easily miss, such as smooth, shiny, rough, colour, etc., as well as interrelationship between different features on the surface. The use of different lighting directions (vertical, oblique, glancing angle) should be explored.

Visual examination can be augmented by close-up photography (Fig. 5.1). Modern high-resolution digital cameras can be useful for this purpose as they eliminate the need for photographic developing and printing, and allow for computer archiving and image processing.

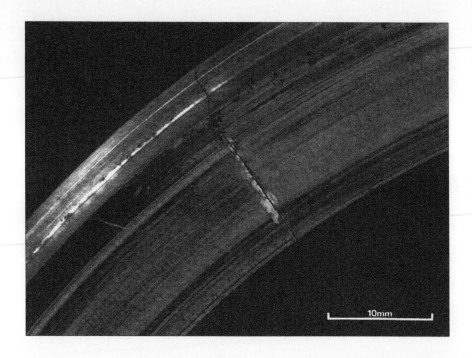

*Fig. 5.1 Close-up photography of the surface of a failed
component (showing a tungsten carbide in cobalt
ring, from a mechanical seal, which has failed by
thermal fatigue)*

A hand-held magnifying glass giving magnifications in the range × 2 to × 5 is an effective addition to direct visual inspection. Also, if the component is small enough to be handled, a binocular microscope with a magnification of up to × 15 is probably the most effective method of all for direct practical surface examination.

If a much more detailed laboratory examination is required optical microscopes with magnifications up to × 1500 are available. Images reveal information about the colour and placement of different features on the surface (Fig. 5.2). One of the limitations of conventional high-powered optical microscopy is the small depth of focus (dependent on microscope objective and thus magnification). This limits the usefulness of conventional microscopes for the examination of rough surfaces and curved components.

Although most optical microscopy is performed using brightfield imaging, additional information can often be obtained under other imaging conditions. Thus crossed polarizing filters can give different contrast depending on the optical activity of the different phases being examined, and differential interference contrast gives contrast that depends sensitively on the topography of the surface being examined (Fig. 5.3).

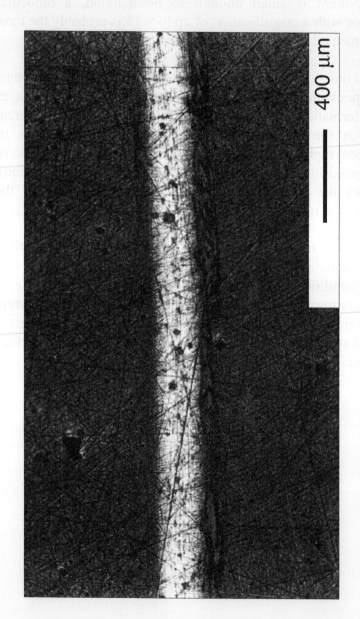

400 μm

(a) Light wear showing build up of titanium oxide at boundaries of wear track

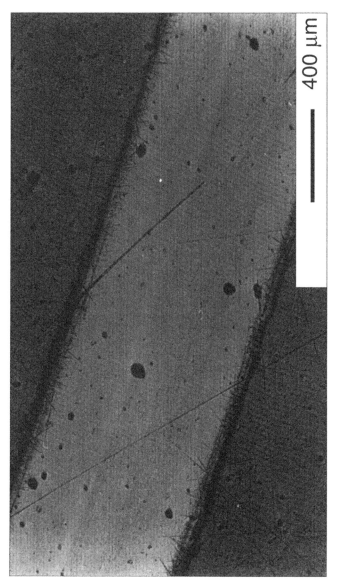

400 µm

(b) *More heavily worn wear track with removal of scratches from centre of wear track by wear process*

Fig. 5.2 *High-magnification images showing surface features. Wear track on titanium nitride-coated samples worn in pin-on-disc configuration (to see this figure parts (a) and (b) in colour, please refer to colour plate section)*

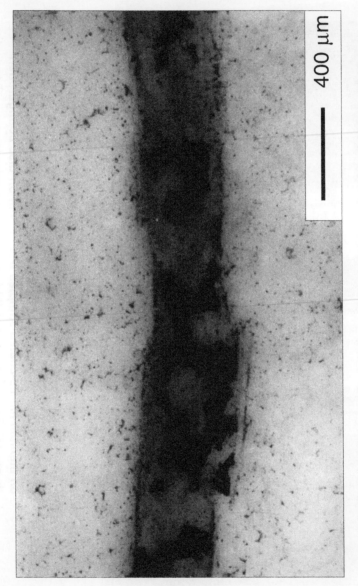

(a) *Imaged with near-crossed polarizer illumination and emphasizing the iron oxide reaction layers on the surface of the wear track*

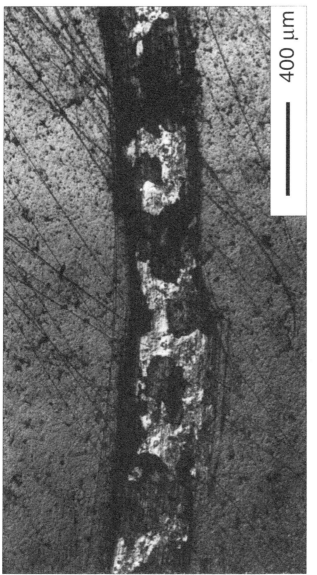

400 μm

(b) **The same area imaged with differential interference contrast emphasizing the transferred layers of steel on the wear track**

Fig. 5.3 **The effect of different lighting arrangements on the images. Iron-rich transfer films on alumina sample formed during wear of steel pin against alumina sample** (to see this figure parts (a) and (b) in colour, please refer to colour plate section)

5.2.2 Electron microscopy

Another very effective laboratory technique for the examination of worn surfaces is the electron microscope. This uses an electron beam interacting with a sensitive screen, rather than a light beam reflected back to the eye.

There are two methods of use for electron microscopes. One, scanning electron microscopy (SEM), scans the surface with an electron beam and the other, transmission electron microscopy (TEM), passes an electron beam through a very thin specimen cut from the surface.

The scanning electron microscope uses secondary electrons that are emitted from the surface of a sample when the electron beam of the microscope hits the surface at the point being imaged. Because the generation of secondary electrons is strongly dependent on the orientation of the surface at the point being examined, the image contains a significant contribution from topographical features on the surface of the sample. The scanning electron microscope can give very high spatial resolution, and has a high depth of focus. However, because the image is formed by an electron beam, sample charging problems can occur for non-conductive samples. These are normally solved by coating samples as required with thin layers of conductive material such as carbon or gold, but for surfaces such as powders or tribological layers where the contact between one part of the surface and another is not very good, charging can still occur.

Another problem with the scanning electron microscope is that it gives poorer detectability for some features on the surface of the sample such as pits. (A classic example of this is the imaging of Vickers hardness indentations in SEM. Even when they are quite large and can be easily detected by optical microscopy or even by eye, it can be very difficult to see the same indentations in SEM.)

Scanning electron microscopy of a surface can, however, reveal details of the surface very clearly at high magnification such as in Fig. 5.4.

Fig. 5.4 *A view with a scanning electron microscope at high magnification. A worn tungsten carbide in cobalt hard metal surface showing deformed and fractured grains*

Scanning electron microscopy is a very effective and useful technique and can give magnifications from 1 to 100 000. It can examine smooth or rough surfaces, flat or curved, and can accept specimens up to 150 mm in diameter. It can also provide information on the chemical composition of the surface by *in situ* X-ray analysis. Scanning electron microscopes also usually have an image reversal facility by which surface replica specimens can have their image converted to that of the real surface. Stereo pair photographs can also be taken to enable height and depth size differences to be obtained by using a stereo mapping device. This can then be compared with the information obtained from surface topography measuring devices.

Transmission electron microscopy is a useful technique where a very detailed analysis of the small-scale structure of surface layers can be made. For some types of samples, near atomic resolution can be achieved.

The main drawback with TEM is that it requires the use of electron-transparent samples. These need to be less than about 200 nm in thickness, and can be very difficult to prepare for solid samples. Normal techniques used for sample preparation are mechanical grinding followed by ion-beam thinning.

One area where sample preparation is extremely easy is in the examination of wear debris. The preparation of samples consists simply of scattering the wear debris on to carbon films and shaking off the loose powder. Sufficient particles adhere to the carbon film to allow examination and analysis at high magnification.

5.3 Topographical information including surface finish

The shape of a worn surface gives clues about how it was formed. For instance, the presence of deep grooves on a surface is deeply suggestive that abrasion processes have been occurring; in fact, one definition of abrasion relies on this observation. Although considerable information on the shape of a surface can be given by direct visual examination, and by optical and scanning electron microscopy, actual quantitative information can only be obtained by measuring the response of probes of different types as they move from position to position on the surface.

5.3.1 Mechanical probes

Mechanical probes have been used for many years in the determination of surface topography, and must be the most common technique used for the characterization of surface topography (see Fig. 5.5). Normally, a diamond stylus with a well-defined tip geometry (e.g. tip radius of a few micrometres) is used as the mechanical probe. Either the stylus is scanned over the sample, or the sample is scanned under the probe. Variations in the height of the surface from point to point result in a movement of the probe which is recorded to give a measure of the sample height at that point. The geometry of the tip is well defined by national and international standards. One of the potential drawbacks with stylus instruments is that the movement of the stylus over the surface can cause damage to the surface when soft materials are being examined.

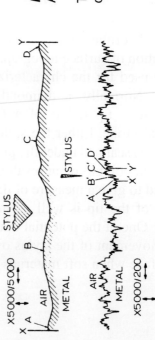

Instrument construction

Stylus T measures the surface profile relative to reference surface P followed by stylus S

Instrument output

Actual surface profile

The recorded output has a compressed horizontal scale

Fig. 5.5 Surface topography measurement with a mechanical probe

5.3.2 Optical probes

Optical probes are becoming increasingly popular. Their advantages are that they are non-contact devices, which cannot damage the examined surface, and for full-field instruments, they can be much faster than mechanical systems.

There are four main types of optical instruments:

1. The first comprise systems based on interference microscopy. The basic principle is that an interference pattern is formed by the combination of light reflected from the surface with a reference beam. Often the sample is scanned in height while several patterns are captured (sometimes continuously) and analysed to calculate surface height information. These instruments have good spatial and height resolution, and have high measurement rates. There are some concerns that errors occur when rough surfaces are measured due to the difficulty in analysing the very complex interference patterns that result. These instruments also normally measure a single field of view at a time, limiting the area that can be examined.

2. The second type of instrument is the confocal microscope. The utilization of narrow apertures in the optical system gives optical sectioning, where light is only captured by the microscope from the plane in focus intersecting the sample. The height of the sample is then scanned, and a set of optical sections is collected and analysed to give surface height data. The instrument gives good spatial and height resolution, but is again normally limited to single fields of view. A particular advantage of this technique is that images of the examined area are acquired simultaneously with the height information.

3. The third type of optical system is based on compact disc (CD) reading technology. This has been used as the basis of optical non-contact surface topography probes. They normally use infrared lasers and gain the height information through a servo feedback mechanism which monitors the size of the spot focused

from the reflected light. These systems have the advantage that large samples can be examined, but the large spot size limits the spatial resolution of the device.

4. The fourth type of instrument also uses laser probes, but relies on a triangulation system to calculate the surface height.

Chapter 6

A SUMMARY OF THE APPROACH TO WEAR TESTING

The previous chapters have summarized the nature of industrial wear problems and have given some guidance on their solution, including the selection and operation of wear tests on materials, to provide additional guidance. Since there are many types of machines, wear mechanisms, and industrial environments, the overall pattern of problems can be fairly complex. This chapter, therefore, gives a summary of the suggested approach to solving these problems, that has been described in more detail in the previous chapters of the book.

6.1   Identification of a wear problem

Since some degree of wear is a relatively common occurrence on most machines, it is necessary to determine, for a particular case, whether some remedial action is necessary. Typical factors used in determining whether such action is necessary are:

1. The chances that the wear might result in an accident involving personal injury.

2. The risk that an interruption in the supply of material or services by the machine could cause major customer dissatisfaction.

3. The direct financial losses involved. An average figure for the financial losses incurred by UK industry in 1997, as a result of wear, was 0.25 percent of turnover. This could be used as an indicator for the allowable cost of wear, on a particular machine, as a percentage of its output value.

6.2   Identification of the dominant wear mechanism

If the wear problem is considered significant the next step is to identify the type of wear that is occurring. This may be indicated by the operating conditions of the machine if, for example, it is handling abrasive materials, the mechanism is likely to be abrasive wear. In every case, however, the wear mechanism needs to be confirmed by a detailed examination of the component surfaces (see Sections 2.1 and 2.2).

6.3   Deciding on the action to be taken

The overall position needs to be reviewed to provide guidance on the possible actions that may need to be taken such as:

1. Ameliorate the operating conditions, if this can be done without losing the efficient use of the equipment.

2. Change the material of the affected components, if the existing material is considered non-optimum for its resistance to the wear mechanism that has been identified (see Section 2.4).

3. If there is uncertainty about the choice of a new material, or if the potential cost of using an unsuitable new material is high, then an appropriate wear test should be selected and carried out on a range of possible materials.

4. An understanding of the wear problem may also indicate some design changes that could be made to the machine with the minimum risk of any undesirable side effects.

6.4 Choosing an appropriate wear test

When choosing an appropriate wear test the main factors that need to be taken into account are the wear mechanism, the type of contact condition, and the test conditions that can be reproduced, to represent the actual practical situation (see Section 3.3 and Appendix A).

It will be necessary to carry out a preliminary test on the selected wear machine to check that it reproduces the same wear mechanism and test specimen condition that occurs in the practical application.

6.5 Planning an appropriate series of tests

A number of tests will need to be carried out both to test a number of possible candidate materials, and to allow for some variation in the test conditions for each of them, in order to expose any sensitivity of their test results to relatively minor changes. This should ensure that the test results are representative of the actual operation conditions.

6.6   Carrying out the wear test programme

It will generally be the optimum approach initially to use an organization with wear test equipment and experience to carry out the tests. If these tests result in a successful outcome, and it is anticipated that an ongoing programme for similar tests may be necessary, it may then be considered desirable to purchase an appropriate wear test machine from a reputable supplier. The necessary in-house testing skills and experience can then be developed (see Section 3.4 and Appendix B).

6.7   Assessing the results

The results of tests on alternative materials need to be assessed in terms of the relative wear rates obtained and the surface condition of the specimens. The cost of any potential new materials used for the manufacture of replacement components also needs to be assessed and related to the likely cost savings. It is also possible that some related minor design changes may emerge as being desirable.

Appendix A

DESCRIPTION OF RECOMMENDED TESTS AND THEIR OPERATING CONDITIONS

This appendix contains descriptions of the various wear tests that may be suitable for reproducing wear conditions similar to those occurring in industrial plant and machinery.

Information is also included on the standard test conditions that are recommended for each test in the published standard related to the test technique. These are usually the test conditions that would have been used in any published results for the wear performance of various types of material in these tests.

To investigate an industrial wear problem, these standard test conditions should only be taken as an indicative starting point. In tests relevant to these industrial problems, it is important to use test conditions that are as close as possible to those in the actual application, with the specific objective of producing a pattern of wear on the test specimens that matches the wear pattern on the real component.

The following test methods are included:

Abrasive wear from moving contact with hard, sharp, granular materials

1. Fixed abrasive – abrasive paper or grinding wheel (ASTM G132 gives some information).
2. Rubber wheel, dry abrasive, ASTM G65.
3. Steel wheel, wet abrasive slurry, ASTM B611.
4. Rubber wheel, wet abrasive slurry, ASTM G105.

Abrasive wear from hard, sharp particles trapped between moving surfaces

5. Loose slurry abrasive testing (Miller ASTM G75 is an example).
6. Ball cratering test.

Adhesive wear from the rubbing together of relatively smooth surfaces

7. Sliding wear – uni-directional motion, pin-on-disc, ASTM G99.
8. Sliding wear – reciprocating motion, ASTM G133.
9. Sliding wear – thrust washer test.

Fretting

10. Fretting test system.

Cavitation erosion

11. Cavitation erosion test system, ASTM G32.

Particle erosion

12. Liquid jet erosion test.
13. Gas blast erosion test, ASTM G76.

In these tests, if the test conditions are standardized, they are listed. Otherwise, typical ranges of test conditions are given.

Note: ASTM Standards can be obtained in the UK from:

American Technical Publishers
27/29 Knowl Piece
Wilbury Way
Hitchin
Herts, UK

Tel: +44 (0)1462 437933
Fax: +44 (0)1462 433678

1 Fixed abrasive – abrasive paper or grinding wheel

The simplest types of two-body test systems in concept are those which utilize abrasive papers or abrasive grinding wheels, with a test sample moved with an applied load against the abrasive medium.

A major concern with this type of test is the degradation of the abrasive medium as the test sample is repeatedly moved against it. This has been recognized as a concern, and several strategies have been devised to move the test sample in a complex manner such that it only passes a single point on the abrasive medium once during a test. An example of this is the test devised by Krushchov where the test sample is moved in a spiral over abrasive paper as it is rotated in a similar manner to a long-playing record.

Test conditions

Load; normally in the range 5–100 N

Speed; normally up to 1 m/s

Duration; few tens of minutes

Abrasive; alumina silica, silicon carbide, diamond (abrasive papers or grinding wheel)

Measurements made

Volume of wear (measured directly by profilometry or calculated from mass loss and density measurements)

Examination of worn surface

Friction

Wear displacement (progressive movement of two samples together during wear)

2   Rubber wheel, dry abrasive, ASTM G65

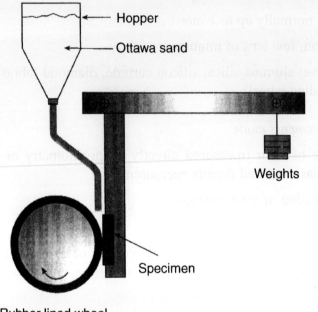

Hopper

Ottawa sand

Weights

Specimen

Rubber lined wheel

ASTM G65 test system

The ASTM G65 test uses a rubber-rimmed wheel as the bed for silica abrasive that is fed from a hopper by a nozzle between the sample and the wheel. The sample is pressed into the wheel by a dead-weight loaded lever. The test is run for a set period and the wear to the sample measured by measuring the volume of material lost through mass loss and density measurements.

The flow of sand into the gap between the surfaces is controlled by the geometry of the nozzle that must be very carefully defined to obtain the required sand flow rate. The abrasive only passes through the wear interface once and is thrown away at the end of a test.

Developments of this standard test by placing the specimen on top of the wheel and feeding the abrasive, under controlled conditions, to the rising upper side of the wheel, give more consistent results. The properties of the rubber lining of the wheel also have a major effect on the results and need to be carefully controlled.

Test conditions as described in the Standard

Abrasive; AFS 50–70 test sand (about 200 μm in size)

Abrasive flow rate; 300–400 g/min

Test load; 45 or 130 N

Number of revolutions; 100, 1000, 2000, or 6000

Wheel speed; 200 r/min

Wheel diameter; 228 mm

Measurements made

Volume of wear

Examination of worn surface

3   Steel wheel, wet abrasive slurry, ASTM B611

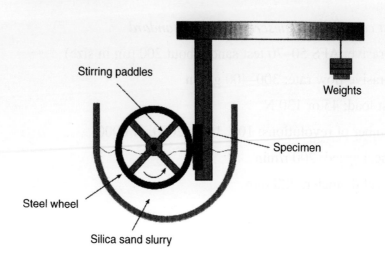

ASTM B611 test system

In the ASTM B611 test for hard metals used in mining applications, a rotating steel wheel is used as the bed for the abrasive that is carried up, out of a bath of slurry surrounding the wheel and test sample, into the wear interface. The abrasive stirring and lifting needs to be observed during the test to ensure that the operation is consistent. The abrasive specified in the test standard is alumina about 600 μm in size. This is industrially unrealistic since the alumina abrasive is very hard (about 2300 HV30) and rarely encountered in practice. A better choice of abrasive from this respect would be silica. The volume of wear is calculated from mass loss and density measurements.

Test conditions as described in the Standard

Test load; 20 N

Duration; 1000 revolutions of the wheel

Abrasive; alumina slurry, 30 mesh grit in proportion of 4 g to 1 cm^3 of water

Wheel speed; 100 r/min

Wheel diameter; 165 mm

Measurements made

Volume of wear

Examination of worn surface

4   Rubber wheel, wet abrasive slurry, ASTM G105

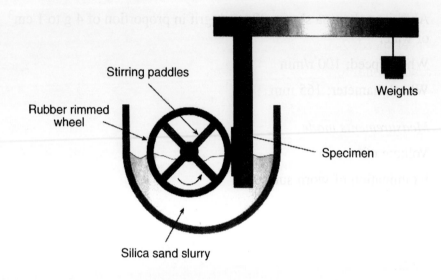

Stirring paddles

Rubber rimmed
wheel

Weights

Specimen

Silica sand slurry

ASTM G105 test system

This test uses a very similar test geometry to the ASTM B611 test
system but the steel wheel is replaced with a rubber-rimmed wheel.
In contrast to the ASTM G65 test the abrasive is carried up, out of a
bath of slurry surrounding the wheel and test sample, into the wear
interface. Straight paddles are placed on the sides of the wheel to
ensure that the abrasive slurry is continually mixed. The test
operation needs to be observed to ensure that adequate mixing is
occurring. It should be noted that the abrasive may pass through the
wear interface more than once, as it is contained in a closed bath and
recirculated through the wear interface during the test.

The volume of wear is measured by mass loss and density
measurements.

Test conditions as described in the Standard

Test load; 222 N

Number of revolutions; 5000

Wheel diameter; 178 mm

Wheel speed; 245 r/min

Rubber condition; Shore A Durometer hardness 50, 60, 70

Measurements made

Volume of wear (calculated from mass loss and density measurements)

Examination of worn surface

5 Loose slurry abrasive testing, ASTM G75

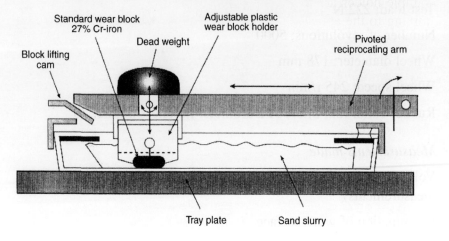

Miller abrasivity test, ASTM G75
(to see this figure in colour, please refer to colour plate section)

There are several loose slurry abrasive test systems that have been used. These include test systems where a sample is pressed against a bed of abrasive material (or vice versa) which can either be dry or damp. It is also quite common to use an adapted polishing machine where a sample is pressed into a dish of abrasive slurry. There is also a reciprocating test with a set of samples moved under a dead-weight loading against an abrasive slurry contained in a tray. This test is standardized as the Miller abrasivity test ASTM G75 as a test of the abrasivity of abrasives; it is also commonly used as an abrasion test.

A major concern with tests of this type is to ensure that the abrasive bed that is achieved in the test, is realistic and well controlled. Particularly for the tests which use modified polishing wheels, there can be difficulties in maintaining a good film of abrasive slurry between the test sample and the supporting dish. This is especially difficult if large abrasives need to be used, and careful consideration needs to be given to strategies for mixing the abrasive slurry as the test proceeds. This is achieved in the Struers micro-abrasion test which uses a semi-automatic polishing system. Three samples are mounted in a holder that is positioned eccentrically from the axis of rotation of the dish containing the abrasive slurry. The dish and the sample holder are then rotated in opposite directions, giving good mixing to the abrasive slurry. This is the system illustrated in Fig. 3.3(b) of this guide.

Test conditions

Abrasive; silica, silicon carbide, alumina or other abrasive

Load; often in the range 5–100 N

Relative speed; up to about 0.5 m/s

Abrasive loading; up to 50 percent by weight loading (slurry needs to be of such a consistency that it can be pumped like a liquid)

Test duration; up to 6 hours

Measurements made

Volume of wear

Examination of worn surface

6   Ball cratering test

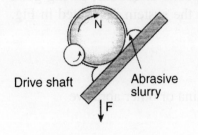

Drive shaft   Abrasive slurry

Rotating polished steel ball rubbing on a test surface, in the presence of an abrasive paste producing a circular crater

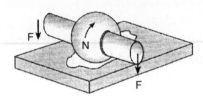

Rotating end-clamped ball rubbing on a test surface in the presence of an abrasive paste to produce a circular crater

Schematic diagram of two different types of ball cratering equipment

Ball cratering is a miniaturized abrasion test method that has been developed over the last few years from two earlier techniques. These were the production of transmission electron microscope samples by dimpling with a wheel or ball to produce a spherical depression that was more easily thinned; the second technique was the use of a ball and appropriate grinding media to produce a spherical crater with known geometry, which cuts through a coating. From measurements of the diameter of the crater in the substrate and the coating, the thickness can be readily calculated.

Through better control of test parameters such as load, speed, abrasive type, suspension, etc., these techniques have been developed readily into a wear test. There is now considerable scope for the use of the test for thin, hard coatings such as TiN, polymeric films such as paints, and other monolithic materials.

One of the key points with this technique is that it enables a test to be carried out on a small area (about 0.5–1 mm across), with little damage to a component – which can remain in service in many cases. The test can easily be made portable, and taken to the component rather than vice versa.

Test conditions

Test load; normally less than 0.5 N but can be up to 20 N in special cases

Abrasive; alumina, silicon carbide silica, diamond, small grit size, normally 4 µm or less

Abrasive loading; 20 percent by volume or less

Ball size; normally about 25 mm

Ball speed; up to 200 r/min

Test duration; up to 5000 revolutions

Measurements made

Size of wear scar

Volume of wear (measured directly by profilometry or calculated from size of wear scar)

Examination of worn surface

Wear displacement (progressive movement of ball into sample)

7   Sliding wear – uni-directional motion, pin-on-disc, ASTM G99

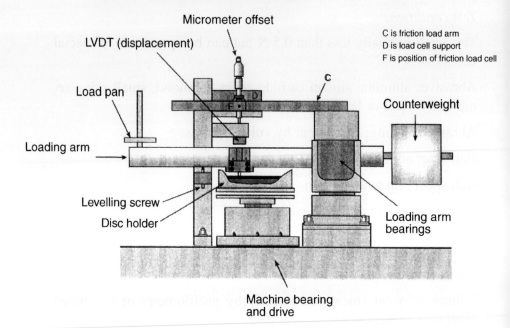

Micrometer offset

LVDT (displacement)

C is friction load arm
D is load cell support
F is position of friction load cell

Load pan

Counterweight

Loading arm

Levelling screw

Disc holder

Loading arm bearings

Machine bearing and drive

Schematic diagram of pin-on-disc test system

The pin-on-disc test has been very widely used. A stationary pin or ball is pressed against a rotating disc. Friction is often measured by recording the force needed to restrain the pin or ball in the tangential direction. Wear displacement (the movement of the samples into one another due to wear of one or other sample) is often measured by use of a linear displacement transducer.

Wear can also be measured in several other ways after the test has been completed. These are:

(a) mass loss measurements combined with a knowledge of the density of the samples;

(b) change in linear dimension of samples;

(c) optical measurement of the size of wear scars (particularly useful for samples with a rounded surface); and

(d) profilometric measurement of wear volume.

Most test systems are only intended for tests under ambient conditions, but test systems have sometimes been modified for use at high temperatures or in different environments.

Test conditions

Test load; up to 10 000 N (in different test systems to cover different load ranges)

Speed; 0.001–10 m/s

Contact geometry; flat-ended pin, chamfered pin, or rounded pin or ball (10 mm diameter is common dimension) against flat

Test duration; 10^3–10^6 s

Test conditions commonly used for validation

Load; 10 N

Speed; 0.1 m/s

10 mm diameter ball

Test duration; 10^4 s

Measurements made

Volume of wear (measured directly by profilometry or calculated from mass loss and density measurements or calculated from size of wear scar)

Examination of worn surface

Friction

Wear displacement (progressive movement of two samples together during wear)

8   Sliding wear – reciprocating motion, ASTM G133

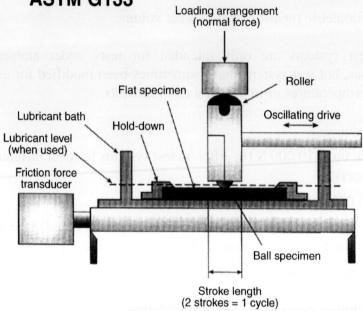

Typical reciprocating test system, ASTM G133

The reciprocating test system has also been very widely used. A stationary pin or ball is pressed against a flat sample that is moved backwards and forwards in a reciprocating motion. There are two types of motion that have been used with reciprocating tests. These are a sinusoidal movement, often generated through some type of cam, and a linear motion, generated by linear servo systems.

Friction is often measured by measuring the force needed to restrain the pin or ball against the direction of travel. Wear displacement (the movement of the samples into one another due to wear of one or other sample) is often measured by use of a linear displacement transducer.

Wear can also be measured in several other ways after the test has been completed. These are:

(a) mass loss measurements combined with a knowledge of the density of the samples;

(b) change in linear dimension of samples;

(c) optical measurement of the size of wear scars (particularly useful for samples with a rounded surface);

(d) profilometric measurement of wear volume.

Most test systems are only intended for tests under ambient conditions, but test systems have sometimes been modified for use at high temperatures or different environments.

Test conditions

Test load; up to 10 000 N (in different test systems to cover different load ranges)

Frequency; 0.1–50 Hz

Stroke; 0.25–50 mm

Contact geometry; flat-ended pin, chamfered pin, or rounded pin or ball (10 mm diameter is common dimension) against flat, or part of piston ring and liner

Test duration; 10^3–10^6 s

Standard ASTM G133 test conditions

Load; 25 N

Frequency; 5 Hz

Stroke; 10 mm

Pin with 4.76 mm radius end

Test duration; 16 min 40 s

Measurements made

Volume of wear (measured directly by profilometry or calculated from mass loss and density measurements or calculated from size of wear scar)

Examination of worn surface

Friction

Wear displacement (progressive movement of two samples together during wear)

9 Sliding wear – thrust washer test

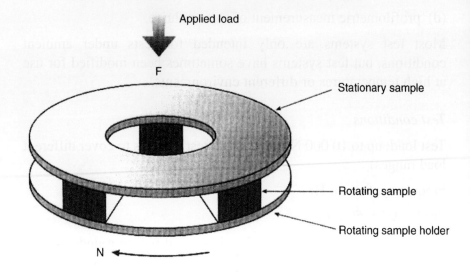

Applied load

F

Stationary sample

Rotating sample

Rotating sample holder

N

Thrust washer test configuration

The thrust washer test is used where testing of conformal large-area contacts is required. Obtaining conformal contact is helped by the large area of the test sample, but it is also often necessary to incorporate additional alignment devices to ensure full conformal contact. Because of the relatively large area of the sample, contact pressures are often lower than in other sliding wear tests such as the reciprocating or pin-on-disc tests.

The two test specimens are often of similar dimensions, but there are advantages in using specimens of different sizes since lateral alignment requirements can be relaxed. In this case the softer sample is made smaller than the harder one and may be divided into individual blocks by incorporating radial grooves in the surface. Friction is often measured by recording the torque needed to restrain the rotation of the stationary sample. Wear displacement (the movement of the samples towards one another due to wear of one or other sample) is often measured by the use of a linear displacement transducer.

Wear can also be measured by mass loss measurements combined with a knowledge of the density of the samples, and for tests with dissimilar sample sizes, by the measurement of the volume of wear in the larger test sample by profilometry.

Most test systems are only intended for tests under ambient conditions, but test systems can be developed for use at high temperatures or in different environments.

Test conditions

Test load; up to 10 000 N (in different test systems to cover different load ranges)

Speed; 0.1–10 m/s

Contact geometry; flat face of ring against flat face of ring

Test duration; 10^3–10^6 s

Measurements made

Friction

Examination of worn surface

Wear

10   Fretting test system

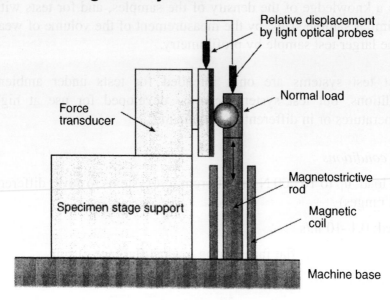

Typical fretting test system

Fretting takes place where small-amplitude motion occurs between loaded contacts. In contrast to other types of sliding wear, the amplitude of motion in fretting is so small that there is always some contact between the same high spots on the two samples. A practical limit on the amplitude of motion in fretting is often taken to be less than 250 μm.

Although the type of motion required is similar to that in reciprocating testing, the small amplitude of motion necessitates considerable care in the design of the test system so that the movement at the point of contact between the samples is correct. This is because inadequate design, particularly with compliant test systems can lead to reduced movement at the point of contact compared with the design value, leading to erroneous results.

Test conditions

Frequency; up to several kHz

Stroke; up to 0.25 mm

Test load; typically in range 2–100 N

Contact geometry; flat-ended pin; chamfered pin, or rounded pin or ball (10 mm diameter is common dimension) against flat

Test duration; up to 10^6 cycles

Measurements made

Examination of worn surface

Wear and surface condition

11   Cavitation erosion test system

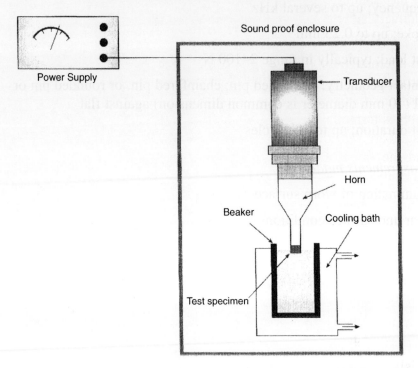

Schematic diagram of vibratory apparatus for cavitation erosion

The resistance of materials to cavitation erosion can be evaluated by two standard tests. These are:

- Cavitating liquid jet, ASTM G134
- Cavitation erosion using vibratory apparatus, ASTM G32

The cavitating jet test system uses a submerged cavitating jet which impinges on a test sample such that cavities collapse on it. This test has the advantage that cavitation is generated in a flowing system so that both the jet velocity and the downstream pressure (which causes the bubble collapse) can be varied independently. However, the test system is quite complex to build and operate, and will not be considered further.

Testing cavitation erosion with the ultrasonic test method is very simple in principle. A test sample is firmly fixed to the end of an ultrasonic horn and is then immersed in an appropriate test liquid. The horn is switched on for a controlled period and the volume of wear is calculated from the mass loss of the sample and a knowledge of the sample density. The test method offers a small-scale, relatively simple and controllable test that can be used to compare the cavitation erosion resistance of different materials.

An alternative approach that uses the same basic apparatus, and is particularly suitable for coatings, is to place the sample in a fixed position at a known distance from the tip of the horn. Cavitating bubbles formed in the liquid then act on the sample. In this case the results are very dependent on the stand-off distance between the tip and the sample, and this parameter needs to be controlled very carefully. The horn and the sample also need to be aligned parallel, otherwise uneven wear can take place.

In both types of test arrangement, the test results are very sensitive to vibratory amplitude of the horn; this needs careful measurement and adjustment to obtain repeatable and reproducible results.

These tests are the only ones in common use in wear testing where wear is not caused by the contact between two solid materials. The wear damage to the surface is caused by the very high localized stresses imposed on the surface when gas bubbles (cavities) in the liquid collapse.

It is also important to control the temperature of the liquid to the specified value. This is normally achieved by means of a cooling bath around the test container or a cooling coil with suitable thermostatic control.

Test conditions (for vibratory apparatus)

Test liquid; normally de-ionized or distilled water to a depth of at least 100 mm

Temperature; 25 °C

Vibratory amplitude; 50 µm peak-to-peak

Frequency; 20 kHz

Measurements made

Wear volume

Examination of worn surface

12 Liquid jet erosion test

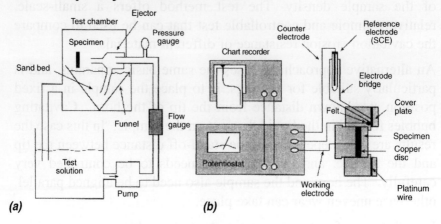

Typical fluid jet test system:
(a) test system; (b) electrochemical measurement cell

A major source of industrial wear problems is through the erosive wear of components from hard particles carried in a jet of fluid. There are currently no recognized standards for this type of test, but there has nevertheless been work carried out in this area.

A typical test system is shown. A volume of fluid is circulated around a closed loop by a pumping system to supply the fluid jet. The jet carries the appropriate concentration of erodent. The sample is placed at the appropriate angle and distance from the nozzle, which sends the jet of fluid containing the erodent on to the sample. The fluid is caught and recirculated.

The key parameters that need careful control in this test are the placement (angle and position) of the sample, the shape of the fluid jet (defined by the nozzle), the fluid pressure, the abrasive loading, and the abrasive shape and size.

Although wear takes place through the abrasive action of the abrasive particles, a major effect also comes through any corrosion that is controlled by the chemistry of the fluid. Thus small variations in the chemical composition such as the pH of the fluid (if water based) can cause major changes in the rates of material loss from the sample, sometimes with dramatic increases in the rate of material loss through synergy between the abrasion and the corrosion.

These corrosion effects need to be carefully controlled. One of the methods that has been used in this respect is to control the electrical potential that is applied between the fluid (via the nozzle) and the sample.

Test conditions

Jet impact velocity; 2–30 m/s

Fluid; normally water but other fluids possible

Impact angle; 20–90°

Erodent; often silica, but other erodents possible

Duration; several tens of minutes

Measurements made

Volume of wear

Examination of worn surface

13 Gas blast erosion test, ASTM G76

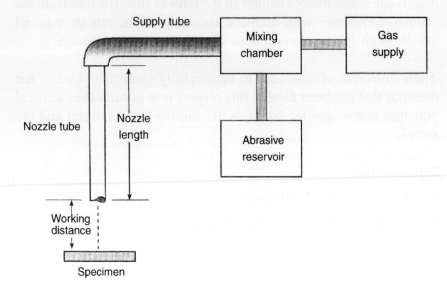

Schematic diagram of gas blast erosion test system, ASTM G76

The gas blast erosion test system uses a stream of high-pressure gas to accelerate a stream of particles through a nozzle towards a test sample. The erodent is only used once before being discarded. The sample is held at a defined distance and angle to the nozzle for the test system.

The most important test parameter, apart from the working distance and angle of impact, is the particle velocity, because the wear increases in line with a power law relationship with velocity. The wear obtained is also dependent on the nozzle geometry (diameter, length, and shape) and internal surface finish of the nozzle. The recommended nozzle size in the ASTM G76 test is 1.5 mm diameter and 50 mm long. This small jet diameter results in the particles 'drilling' into the surface. The deeper the penetration, the more the physical state of the erosion jet will change (due to interaction of the particles hitting the surface and rebounding). These severe test conditions are not appropriate for coatings, for which the test must be designed to determine the wear of the coating, not the substrate material. For some tests it may be desirable to try larger diameter, longer nozzles, in order to obtain test conditions more representative of the practical situation. With a new nozzle it may be necessary to 'run-in' the surface as the exit geometry and surface finish will vary as they are eroded by the particles. The nozzle will, however, quickly settle into a steady state, at which point the particle velocity should be calibrated.

Test conditions

Velocity of erodent; 30 ± 2 m/s

Erodent material; nominally 50 μm angular alumina

Angle of incidence; 90°

Sample stand-off distance; 10 ± 1 mm

Nozzle dimensions; 1.5 ± 0.075 mm inner diameter, at least 50 mm long

Test duration; 10 minutes

Abrasive feed rate; 2 ± 0.5 g/min

Measurements made

Volume of wear

Examination of worn surface

Appendix B

RECOMMENDED DETAILED TEST PROCEDURES FOR THE VARIOUS TESTS

Following the selection of the type of wear test that is expected to give a reasonable simulation of a practical problem, it is important to carry out the actual testing in a manner likely to give reliable and repeatable results.

In many cases the appropriate test procedures are included in the test specifications published by standards organizations, such as the American Society for Testing and Materials (ASTM). There has, however, been a programme of test improvement, carried out by the

National Physical Laboratory, Centre for Materials Measurement and Technology, and guidance on improved test procedures is available from this programme. In addition, there has been a series of international interlaboratory tests on the wear of thin, hard coatings, as a part of the FASTE project funded by the European Union. This appendix summarizes recommended test procedures, which include the results of these various programmes.

The test procedures covered by these recommendations relate to tests for wear between balls/pins and rotating discs, and between balls/pins and plates with relative reciprocating motion.

In each case the procedures cover:

1. important features of the test apparatus that need to be checked;
2. the preparation of the test specimens;
3. the measurement methods to be used during the tests;
4. the carrying out of the test programme;
5. the results which should be reported.

These are described in more detail in the following Sub-sections 1 to 5.

1　Apparatus

Loading system

The system used to load the pin or ball against the disc or plate can be a load beam with spring actuation or dead-weight loading (with or without a lever arm) or can be pneumatic or hydraulic.

- With dead-weight loading, any frictional forces in bearings in the loading system must be negligible in comparison with the loads used in testing. This must be checked at regular intervals or when the loading system is altered. The masses which are used to apply the load must be calibrated and traceable to the National Standard of mass.

- For pneumatic or hydraulic loading, the applied load must be calibrated and traceable to the National Standard of force. The loading system must be designed so that the applied load does not drift by more than 1 percent during the test, and so that variation in loading is not produced by movement of the specimens (e.g. by stiction for pneumatic cylinders).

- With beam and spring, the stiffness of the system will be constant over the entire load range and therefore reduce the effects of load dynamics encountered with dead-weights. The main drawback is that in the absence of automated compensation the load recedes if there is significant wear in the contact.

Specimen mounting and alignment

The specimens should be mechanically clamped in position. The discs or plates should be aligned so that any variation in the surface height along the projected wear track does not exceed 10 μm. This is to ensure that there is no variation in the applied loading arising from the inertia of the components.

In the case of pin/ball-on-disc tests, various orientations have been used, with the disc surface horizontal or vertical, and the pin/ball above or below the disc. However, the most usual arrangement is to have the disc surface horizontal, with the pin/ball pressed against the top surface of the disc.

Specimen test environment

The specimen holders should be designed to give the shortest possible thermal path to a sink of heat, such as the bulk of the test system, so that the thermal environment will be constant from test to test. In some cases, forced cooling may be required to control the contact temperature. For thin, hard coatings the temperature of the test does not have a major effect on test results. Tests can be carried out under ambient conditions within the range 15–25 °C. However, the test temperature should be recorded. The humidity of the air surrounding all tests should be controlled to 50 ± 5 percent RH (relative humidity) unless some other humidity has been specified.

Driving system

For pin/ball-on-rotating-disc tests the motor drive must give a constant speed of rotation, and must have sufficient power to continue at a constant speed (to ± 1 percent) under the frictional loading that will be experienced in tests.

For reciprocating pin/ball-on-plate tests the drive system must give a well-controlled stroke profile, and must have sufficient power to maintain this profile under the frictional loading that will be experienced in tests. The stroke profile should be checked at periodic intervals. Two stroke profiles that are commonly used are a sinusoidal variation in displacement and speed (normally achieved through crank and similar drives), and a square wave motion (normally achieved through a linear drive system). Either the plate or the ball or pin can be driven; commercial test systems are available for either. However, it should be noted that due to the increased mass of the plate sample and its holder, the maximum frequency that can be obtained with a system which moves the plate is considerably smaller than when the ball or pin is moved.

Machine dynamics

The test machine should be designed so that the natural frequencies of the testing system are much higher than any vibrations generated from sources such as the motor drive, disc rotation, or plate reciprocation.

The specimen support must also be very rigid, particularly in the direction in which it is rubbed, in order to avoid stick–slip vibration occurring. The stiffness of the loading beam is usually the most critical factor.

2 Test specimens

Materials

The usefulness of the wear testing is increased if the materials under test are well characterized; conversely, if there is little materials information the wear test can never be compared with results of other tests. A minimum description of the material would include the chemical composition, the condition of the material (e.g. the heat treatment for materials such as steel), the hardness of the material, and a micrograph of the surface. Other information which would certainly be useful could be obtained from surface analysis, X-ray diffraction, examination of the microstructure of the material, and surface texture analysis. If the materials conform to any national or international specifications or standards, this should be stated.

Dimensions and surface condition

The contact dimensions of pins/balls are usually much smaller than the contact area of their mating disc or plate, so that thermal and contact duration conditions are different for the two components in contact.

An important parameter, therefore, is the overlap ratio, which is defined as the ratio of the area on the disc or plate that is in contact with the pin at some stage, to the area on the pin that is in contact with the disc or plate all the time. Specimen sizes which have been used in recent interlaboratory exercises are discs 42 mm in diameter or plates 30 mm × 20 mm, both used with balls 10 mm in diameter. Pins 6 mm and 10 mm in diameter are also commonly used specimens. It is recommended that, if pins are used, the end of the pin in contact with the disc should be prepared to a radiused end to give non-conformal contact conditions. This will avoid alignment problems. Typically a spherical radius of 5 mm is used.

The discs or plates should be manufactured with flat and parallel faces. The finish on the specimens will be dictated by the programme of testing, but it is clear that the surface finish has a large effect on the friction and wear results that are obtained, with very rough surfaces leading to high wear, at least in the initial stages of a test. Conversely, for very high-quality finishes which are very flat and smooth, wringing may occur leading to abnormally high friction and possible wear. Care must be taken in the preparation of the surfaces of the specimens to avoid subsurface damage.

For tests on the wear resistance of thin, hard coatings the coating is put on the disc or plate and tested in contact with an alumina ball 10 mm in diameter. Polycrystalline alumina balls (better than 99 percent alumina) are a preferred material for these tests on thin, hard coatings as it has been found in preparatory work that they are essentially inert in conditions of wear against TiN coatings. For other coatings, other ball materials may be preferable. The coated surface should be flat with a surface finish better than 0.05 μm R_a, to avoid difficulties in the measurement of wear volume. The balls should have a polished finish (better than 0.025 μm R_a).

3   Measurement methods

Friction measurement

Friction measurements are made with a load cell, or instrumented deflection beam; these must be calibrated at regular intervals. The coefficient of friction is then calculated as the ratio of this measured frictional load to the applied normal load. If a normal load cell is also monitoring the instantaneous normal load, analogue or digital dividing may be used to derive a true instantaneous coefficient of friction. Otherwise the average normal applied load should be used in the calculation of coefficient of friction.

Consideration should also be given to the mechanical attachment of the friction load cell to both the specimen holder and the machine frame. A common practice is to allow the specimen holder to rest against the measurement point on the friction load cell. Care should then be taken to eliminate friction between the specimen holder and the friction load cell; this may alter the nominal load applied between the test surfaces. It should be recognized that this arrangement gives different dynamic characteristics from the situation where the specimen holder is clamped to the machine frame through the friction load cell.

The friction measurements should be recorded digitally, with details of the sampling rate and any data processing noted. It is recommended that a sampling rate of at least 100 samples per cycle is used, when possible. With computer control of the measurements, the sampling rate can be adjusted to suit the conditions, with a reduced rate when the measurements are steady. It is also possible to arrange for the test to be stopped if the friction exceeds a critical value. This helps to preserve the test specimen for subsequent examination. Computer control also improves the repeatability from test to test, particularly if different operators carry out the tests.

Contact potential measurement

If a voltage of about 50 mV is applied across the wear contact by means of a potential divider, the variation in the voltage can be used to indicate the degree of asperity interaction in the contact.

In lubricated contacts the contact potential can be used as a qualitative measure of the presence of an oil film or a chemical additive film on the contact surfaces.

Wear measurement by specimen displacement

Wear measurement by specimen displacement is normally measured by a transducer which is positioned to record the movement of the pin or ball relative to the starting level of the disc or plate. The output of the transducer is then taken to be equivalent to the wear displacement. This assumes that the unworn level of the disc or plate does not change with time. There is a dependence on the position of the measurement point on the specimen holder. Instantaneous readings may require correction for thermal expansion.

Wear measurement by mass change

The mass change of the specimen is measured as the difference in mass of the specimen recorded before and after the test. The specimens should be cleaned to remove any foreign bodies or loose adherent layers from them before weighing in a controlled way. Any debris collected should be retained for further analysis. The calibration of the balance must be checked at regular (e.g. annual) intervals. It is essential that specimens are handled as little as possible before weighing, and that clean, lint-free gloves or clean tweezers are used to minimize contamination of the specimens.

For mass loss measurements, the wear volume is calculated from the relationship

$$\text{Wear volume} = \frac{M}{\rho}$$

where M is the mass loss and ρ is the density. Note, however, that for the coated samples the volume loss can only be calculated using this formula if the wear is contained entirely within the coating and the coating density is known.

Wear measurement by profilometry

Surface texture instruments should be used to characterize the texture of the surfaces before and after testing; they can also be used to make measurements of wear volumes (or measurements that can be used to calculate wear volumes).

For small areas of wear such as the wear scars on balls, three-dimensional instruments can provide a highly accurate direct measurement. However, if only two-dimensional instruments are available, the volume has to be estimated from the profile trace generated across the scar. Similarly, for wear tracks on the flat, the wear volume has to be estimated from the cross-sectional area of the wear scar at points along the track.

It is recommended that for wear scars on balls at least two perpendicular profiles are used for this estimation (one perpendicular and one parallel to the direction of sliding across diameters of scar). If the wear scar on a ball (or spherically capped pin) is flat, then the wear volume V for the ball can be calculated from the formula

$$V = \frac{1}{3}\pi h^2 (3r - h)$$

where h is the radial wear at the centre of the scar and r is the radius of the ball, or from the approximate formula (accurate to within 1 percent for $d < 0.3r$)

$$V = \frac{\pi d^4}{64r}$$

where d is the wear scar diameter.

Measurement of wear to a disc should be carried out by making at least four perpendicular measurements of the profile across the wear track (spaced equally around the disc and aligned parallel and perpendicular to any lay in the texture of the disc, along radii of the disc).

Measurement of wear to a plate, from a reciprocating test, should be carried out by making at least six perpendicular measurements of the profile across the wear track (spaced equally along the wear track on the plate such that the measurements near to the end of the stroke are $0.1 \times$ the length of the stroke from the ends of the wear track).

Optical wear scar measurement

For balls or spherical-ended pins the size of the wear scar can also be measured by optical methods using projection systems or microscopes. It is important that error in the measurement is not introduced by misalignment of the measurement systems. The particular measurement technique that is used must be calibrated at regular intervals.

4   Testing

Specimen cleaning

Metal and ceramic specimens should be ultrasonically cleaned with acetone or ethyl alcohol for 15 minutes, this should be followed by warm air drying, a final rinse with fresh acetone or alcohol, then drying for 30 minutes in an oven at 110 °C. After cleaning, the specimens should be stored for 24 hours in the same environment that will be used for the testing to allow the sample surface condition to equilibrate with the environment.

Polymeric/plastic materials should also be ultrasonically cleaned in a non-reactive liquid, which leaves no deposit on evaporation. If the application involves operation in water or a very humid environment, the specimens should be cleaned in water and stored under water to ensure consistent water retention.

Specimen installation

The specimens should be clamped firmly into position in the test machine, ensuring that the contacting surfaces of the specimens are not touched. On a rotating disc test the radial position of the pin should be adjusted to give the correct wear track diameter. The vertical run-out of the disc or plate surfaces should be checked at less than 10 μm.

Then, with the specimens out of contact, the motor control should be adjusted to give the correct speed. With the machine stationary the zero of the friction measurement system should then be checked by adjusting the pin/ball holder so that no loading of the friction load cell occurs. The loading system should be adjusted to give the required applied load.

Environmental conditioning

The environmental conditioning system should be started, set at the required condition of 50 percent RH, and sufficient time allowed to equilibrate before the test commences. It is recommended that the pin/ball should be lifted clear of the disc or plate and the motor of the test machine started so that the motion does not create a difference in the environmental conditioning, during the actual test.

Start of the test

The data recording equipment should be set appropriately and started, and the motor drive started, if not already running as a part of the environmental conditioning procedure. When the speed is constant, the pin/ball should be lowered carefully on to the disc or plate to start the test. The rate of application of the load, i.e. time from zero to full load, should be controlled carefully and recorded so that it can be repeated in future tests. Periodic inspection of the test is recommended to ensure that the recording equipment is performing correctly and that the test is proceeding as expected.

Test procedures

Because friction and wear are processes where some variation in results from test to test is a natural consequence of the statistical nature of the contact between the two surfaces, it is recommended that sufficient tests should be carried out under a single set of test conditions to reach the required confidence in the measurements. The number of tests that are needed will depend on the scatter in results that is observed. This recommendation may be relaxed when carrying out friction and wear mapping, for example. The scatter from the observed trend of results then gives information about the uncertainty in the measurements.

It is recommended that the appropriate standardized analysis methods laid down in the ISO standard on accuracy (trueness and precision) of measurement methods and results should be used in the calculation of uncertainty in friction and wear measurements.

In the case of tests on the wear resistance of thin hard coatings, it is recommended that three repeat tests should be carried out under conditions where perforation of the coating does not occur. Satisfaction of this condition should be checked by initial scoping tests. These scoping tests should be carried out by varying load, test duration, and rotational speed, or reciprocating frequency and stroke. The main tests should be carried out with at least three test durations, giving a total of at least nine tests for each coating system.

Test conditions which have been found to be satisfactory for 5 µm TiN coatings are given in Table B1.

Table B1

For rotating disc tests		For reciprocating plate tests	
Load:	10 N	Load:	10 N
Speed:	0.1 m/s	Speed:	0.1 m/s
Wear track diameter:	32 mm	Stroke:	5 mm
Test duration: (one test for each)	5000, 7500, and 10 000 seconds	Frequency:	2 Hz
		Test duration: (three tests for each)	2500, 5000, and 7500 seconds

Test termination

The test should be terminated when the appropriate test duration has been reached.

Examination of specimens and debris after test

After testing, the specimens should be examined and the appearance of the worn surfaces and any debris adhering to the surface noted. The specimens should be cleaned, and profilometry and measurement of wear scars carried out to determine the volume of wear that has occurred. The specimens should be stored in dry, clean conditions.

Analysis of results

The detail of the analysis that will be required will depend on the precise requirements of the testing programme. However, a minimum necessary requirement is the plotting of displacement and friction curves for the test, the calculation of initial, maximum, and steady-state coefficient of friction, and the calculation of wear volumes for the pin and disc from mass loss measurements or from profilometry measurements.

In the case of tests on thin, hard coatings it is recommended that the total energy dissipated during each test should be calculated as

$$\text{TE} \;=\; \Sigma\, v\, \Delta t_i\; W \quad \text{for all friction coefficient data points } f_1$$

where v is the relative speed, Δt_i is the elapsed test time, and W is the applied load.

If the total volume of wear is then plotted against the total energy, TE, a linear relationship should be observed. The slope and intercept of the linear fit to the data points should be calculated by regression analysis.

5　Reporting the results

The following results should be reported and recorded:

1. A description of the test system emphasizing the loading system employed, and the friction measurement system.

2. Details of the specimens including composition, geometry, and surface finish. For a coated sample this should include details for both the substrate and the coating.

3. Details of sample cleaning and any pre-conditioning. Also details of the environmental conditions such as temperature and relative humidity.

4. A table of test results including rubbing distance and speed, test duration, profilometric results, mass loss results, total wear volumes, and wear rates. For thin, hard coating tests, this should also include the total energy dissipated.

5. Graphs of the variation of friction and wear displacement with elapsed time. For tests on thin, hard coatings a graph should be included of total wear volume with total energy dissipated.

6. Files containing data on the variation of friction and wear displacement with time. The format of the data should be described.

7. Any observations on the appearance of the samples after testing should also be reported.

Definitions

The test procedures have included a number of technical definitions and the meanings of each of these are as follows:

Applied load Load applied between two contacting surfaces normal to the tangent of the surfaces.

Instantaneous coefficient of friction The instantaneous value of friction divided by the instantaneous value of applied load. This is often approximated to the instantaneous value of friction divided by the mean applied load.

Cyclic average coefficient of friction for a rotating disc test The average value of the instantaneous coefficient of friction calculated for a single cycle of rotation of the disc.

Cyclic average coefficient of friction for a reciprocating test The average value of the instantaneous coefficient of friction calculated for a single cycle. Two possible ways of calculating this are (see Fig. B1):

• to integrate the size of the friction–horizontal motion friction loop and divide this integrated value by the magnitude of the stroke to arrive at a value of friction coefficient.

• to average the absolute magnitude of the friction, excluding the values of friction towards the end of the stroke where the value may be affected by component inertias towards the ends of the stroke.

Initial coefficient of friction The coefficient of friction at the start of the test, during one cycle or rotation.

Maximum coefficient of friction The maximum coefficient of friction measured during the test. In the absence of transitions in wear behaviour this often occurs shortly after the start of a test. This value is obtained as the mean value over a period in which there are five full cycles of rotation or reciprocation.

Steady-state coefficient of friction The average value (or range of values) of the coefficent of friction over a long period of time in

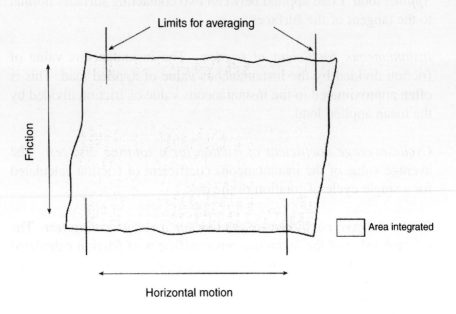

Fig. B1 Calculation of cycle friction

which the coefficient of friction is essentially constant. Such behaviour occurs after the initial transient changes in coefficient of friction have taken place.

Note: In some tests steady-state behaviour does not occur, with coefficient of friction continuing to vary over the duration of the test.

Dimensional wear coefficient, k This is defined by the equation

$$k = \frac{V}{LW}$$

where
   V = volume worn away
   L = total sliding distance
   W = applied load

Wear volume This is the loss of volume to the specimen after a test.

Appendix C

GUIDE TO NOTATION

Table C1

Symbol	Used for
A	Nominal contact area of wearing surface
d	Diameter of wear scar on ball sample
h	Depth of wear
k	Dimensional wear coefficient
k_d	A constant
L	Distance of sliding
M	Mass loss
r	Radius of ball sample
t	Time
v	Velocity of sliding
V	Wear volume
W	Applied load
μ	Coefficient of friction
ρ	Density

Appendix D

BIBLIOGRAPHY FOR FURTHER READING

This guide has given a simple introduction to the different mechanisms of wear and has given guidance on the choice of suitable tests for particular applications. There are a number of books and other documents which are available if more information is required; some sources of guidance, which are by no means exhaustive, are given here.

A useful introduction in the consideration of the framework for friction and wear measurement is given in *Tribology: a Systems Approach to the Science and Technology of Friction, Lubrication and Wear* by H. Czichos (**1**)

General information on tribology is available in several handbooks such as the *Wear Control Handbook* by M. B. Peterson and W. O. Winer (**2**), *The ASM Metals Handbook: Vol. 18, Friction, Lubrication and Wear Technology* (**3**), the *Handbook of Tribology: Materials, Coatings and Surface Treatments* by B. Bhushan and B. K. Gupta (**4**), *Fundamentals of Friction and Wear of Materials*, edited by D. A. Rigney (**5**), *New Directions in Lubrication, Materials, Wear and Surface Interactions* by W. R. Loomis (**6**), *Tribology: Friction and Wear of Engineering Materials* by I. Hutchings (**7**), *Engineering Tribology* by J. Williams (**8**), and *Friction Science and Technology* by P. J. Blau (**9**).

More specific information is given in ASTM Special Technical Publications (STPs) on friction and wear testing of metals (**10**), plastics (**11**), coatings (**12**), elastomers (**13**), ceramics (**14**), and advanced materials (**15**), and an ASTM STP on the fretting fatigue test methods (**16**). Useful information is also available in NPL Report CMMT (A) 92, December 1997, 'Wear testing methods and their relevance to industrial wear problems' (**17**) and in books by Summers-Smith (**18**, **19**).

Compilations of data are available in the *Tribology Handbook* by M. J. Neale (**20**) and the *Elsevier Materials Selector* (**21**).

Many papers on friction and wear testing procedures are given in the proceedings of conferences such as the biennial *International Conference on the Wear of Materials* (**22**), and the peer-reviewed journals *Wear* (**23**), *Tribology International* (**24**), and Institution of Mechanical Engineers Proceedings, *Part J, Journal of Engineering Tribology* (**25**).

1　　**Czichos, H.** (1978) *Tribology: a systems approach to the science and technology of friction, lubrication and wear* (Elsevier).

2　　**Peterson, M. B.** and **Winer, W. O.** (1980) *Wear Control Handbook* (American Society of Mechanical Engineers).

3   ASM (1992) *ASM Metals Handbook, Vol. 18, Friction, Lubrication and Wear Technology* (ASM International).

4   **Bhushan, B.** and **Gupta, B. K.** (1992) *Handbook of Tribology: Materials, Coatings, and Surface Treatments* (McGraw-Hill).

5   **Rigney, D. A.** (1981) *Fundamentals of Friction and Wear of Materials,* papers presented at the 1980 ASM Materials Science Seminar, October 1980, Pittsburgh, Pennsylvania, USA (ASM International).

6   **Loomis, W. R.** (1985) *New Directions in Lubrication, Materials, Wear and Surface Interactions* (Noyes Publications).

7   **Hutchings, I. M.** (1992) *Tribology: Friction and Wear of Engineering Materials* (Edward Arnold).

8   **Williams, J. A.** (1994) *Engineering Tribology* (Oxford Science Publications).

9   **Blau, P. J.** (1995) *Friction Science and Technology* (Marcel Dekker).

10   ASTM (American Society for Testing and Materials) STP 615: *The selection and use of wear tests for metals.*

11   ASTM STP 701: *Wear tests for plastics: selection and use.*

12   ASTM STP 769: *Selection of wear tests for coatings.*

13   ASTM STP 1145: *Wear and friction of elastomers.*

14   ASTM STP 1010: *Wear tests for ceramics.*

15 ASTM STP 1167: *Wear testing of advanced materials*

16 ASTM STP 1159: *Standardisation of fretting fatigue test methods and equipment.*

17 NPL Report CMMT (A) 92: *Wear testing methods and their relevance to industrial wear problems* (1997).

18 **Summers-Smith, J. D.** (1997) *A Tribology Casebook* (Mechanical Engineering Publications, London).

19 **Summers-Smith, J. D.** (1994) *An Introductory Guide to Industrial Tribology* (Mechanical Engineering Publications, London).

20 **Neale, M. J.** (1973, second edition 1995) *Tribology Handbook* (Butterworth–Heinemann).

21 **Waterman, N. A.** and **Ashby, M. F.** (1992) *Elsevier Materials Selector* (Elsevier).

22 Proceedings of the International Conferences on *The Wear of Materials* (1977, 1979, 1981, 1983, 1985, 1987, 1989, 1991) (American Society of Mechanical Engineers).

23 *Wear* (journal published by Elsevier).

24 *Tribology International* (journal published by Butterworth–Heinemann).

25 *Proc. Instn Mech. Engrs, Part J, J. Engng Tribology* (journal published by Professional Engineering Publishing, London).

Index

Printed and bound by CPI Group (UK) Ltd, Croydon, CR0 4YY

08/05/2025

01864831-0001